A Manual for
Primary Human
Cell Culture

A MANUAL FOR PRIMARY HUMAN CELL CULTURE

Jan-Thorsten Schantz
Kee Woei Ng

National University of Singapore

World Scientific

NEW JERSEY • LONDON • SINGAPORE • BEIJING • SHANGHAI • HONG KONG • TAIPEI • CHENNAI

Published by

World Scientific Publishing Co. Pte. Ltd.
5 Toh Tuck Link, Singapore 596224
USA office: 27 Warren Street, Suite 401-402, Hackensack, NJ 07601
UK office: 57 Shelton Street, Covent Garden, London WC2H 9HE

British Library Cataloguing-in-Publication Data
A catalogue record for this book is available from the British Library.

ISBN 981-238-796-X

Typeset by Stallion Press
Email: Sales@stallionpress.com

Printed in Singapore.

This book is dedicated with love to

Annika and Elias.

Peishan and my family.

Preface

As part of the boom in biotechnology, particularly in *Tissue Engineering*, primary human cell culture has become a major pillar in academic research and in the biopharmaceutical industry. Obtaining a viable culture from a tissue sample and maintaining it for experimental, diagnostic or therapeutic purposes can be quite a challenge. Based on laboratory protocols and practical experience from many years of primary cell culture, we provide the basic steps necessary for culturing primary human cells.

Written by students for students, this manual is designed to serve as a practical guide in primary human cell culture. We did not intend to bring out a large compendium and left lots of space for notes and the design of manual is such that it can be continuously upgraded and extended. The content of this manual is by no means exhaustive. Protocols for specific cell types, out of over 200 different cell types in the human body, were selected from major tissue groupings in the body. These should serve as foundations for individual researchers to experiment, explore and establish niche protocols for their specific needs. For practicality, we have chosen a physical format that can fit into the pocket of a lab coat, adapted from the clinical checklists available for residents and trainees in medicine.

We would like to thank all clinical collaborators, especially Associate Professor T.C. Lim, surgeon-in chief of the Division of Plastic Surgery at the National University Hospital, who had provided us with a variety of tissue samples and thus contributed to the development and characterization of several cell culture systems. Special thanks go to Assistant Professor D. W. Hutmacher, from the Tissue Engineering Laboratory at the Division of Bioengineering, for his continuous support and stimulating comments and GIBCO-Invitrogen and NUNC for their support in reviewing the manuscript. Last but not least, we would like to thank Miss Hu Xiaoling & Mr. Kenneth Lim for their help in editing the protocols and all the students who have been trained over the years in cell culture techniques and who have made many valuable suggestions.

J-Th. Schantz & K.W. Ng
March 2004

Foreword

I am honored to prepare this foreword to *A Manual for Primary Human Cell Culture*. Primary human cell culture has become a major element in academia and the rapidly growing biotechnology sector. Human cell culture is being employed not only for basic research purposes, but also for diagnostic and therapeutic applications. Cell transplantation and, in particular, tissue engineering have successfully been established to investigate and to treat a variety of diseases. Researchers and clinicians using those technologies are commonly referring to animal cell culture protocols or protocols that have been established for long-term culture of immortalized cell lines. However, primary human cells are often difficult to isolate and have particular requirements for maintaining them in culture.

Based on the authors' own benchwork practice and teaching in human cell culture techniques, this manual addresses a vital need to provide clear and concise written protocols for scientists and students embarking on primary human cell culture. Thus, it allows the readers to benefit from the advice and experience of scientists actively working in this field. Mastery of the information in this manual should provide the basic elements of primary human cell culture and stimulate trainees to delve deeper into more stratified literature and explanatory cell culture texts. This protocol collection will serve as a valuable reference guide and, by leaving plenty of blank spaces, it encourages the user to modify sections according to one's own personal style. The authors are to be commended for writing this brief, yet very informative, manual that I am confident will prove valuable in current laboratory practice and will find a wide distribution in research and clinical institutions.

Mark A. Randolph MAS
Harvard Medical School
Massachusetts General Hospital

Table of Contents

List of Figures

N.B. Figure numbers correspond to chapter numbers.

List of Contributors

Ziyuan Cheng, M. Eng.
Department of Ophthalmology
National University of Singapore
Lower Kent Ridge Road, 119074 Singapore
Singapore

Amy Chou Ai Mei, B. Sc.
NUS Graduate School
National University of Singapore
MD11, Clinical Research Centre
10 Medical Drive, 117597 Singapore
Singapore

Michaela Endres, M. Sc.
Tissue Engineering Laboratory
Department of Rheumatology
Charité, Tucholsky Str. 2
Humboldt University
10117 Berlin
Germany

Connie Er Poh Nee, M. Sc.
Temasek Laboratories
1 Research Link, National University of Singapore
117604 Singapore
Singapore

Steve Gorfien, Ph.D
GIBCO Cell Culture
Invitrogen Corp
Grand Island, 3175 Stanley Road
NY 14072
USA

Klaus Gossens, M. Sc.
The Biomedical Research Center
2222 Health Science Mall
Vancouver
BC, V6T 1Z3
Canada

Hwei Ling Khor, M. Eng.
Max Planck Institut für Polymerforschung
Ackermann Weg 10
55128 Mainz
Germany

David Leong Tai Wei, B. Eng.
Department of Biological Sciences
National University of Singapore
14 Science Dr. 4, 117543 Singapore
Singapore

Robin Ng Boon Leong, B. Sc.
Tissue Engineering Laboratory
Division of Bioengineering
National University of Singapore
9 Engineering Dr. 1, 117576 Singapore
Singapore

Reida Menshawe El Oakley, MD, FRCS
Gleneagles Medical Center
Sdn Bhd Jerudong Park BG 3122
Brunei Darussalam
Brunei

Arun Singhal, Ph.D.
GIBCO Cell Culture
Invitrogen Corp
Grand Island, 3175 Stanley Road
NY 14072
USA

Xian Wei Wang
Institute of Bioengineering and Nanotechnology
31 Biopolis Way
The Nanos, 138669 Singapore
Singapore

Timothy W.L. Wong, Ph.D.
GIBCO Inivitrogen Hong Kong Limited
Concord Technology Center
98 Texaco Road, Tsuen Wan
Hong Kong S.A.R.

A

Principles of
Biosafety

2 ♦ *Principles of Biosafety*

When working with human tissue samples, an acute awareness of possible risks and a clear concept of biological safety are essential to prevent occupational acquired infections as well as the release of pathogens into the environment.

The laboratory facilities should have restricted access to persons whose presence is required to perform cell culture work and who have been instructed in biosafety principles.

A.1 Categories of Biosafety

Biosafety Level 1 (low risk)
- Laboratories appropriate for undergraduate training and teaching
- Work is done with defined and characterised strains of viable microorganisms not known to cause any disease in healthy adult humans

Safety Requirements: sink for handwashing

Biosafety Level 2 (moderate risk)
- Laboratories appropriate for diagnostics and teaching (graduate and postgraduate level)
- Work is done with agents that are associated with human diseases (microorganisms like: Hep. B, HIV, most bacteria) as well as human body fluids, tissues and primary human cell lines

Safety Requirements:
1. Primary Barriers: face protection, gowns, gloves, Biosafety Class II cabinet
2. Secondary Barriers: sinks for handwashing, waste decontamination facilities

Biosafety Level 3 (moderate-high risk)
- Laboratories appropriate for diagnostics, teaching, research or production facilities
- Work is done with exotic agents with a potential of respiratory transmission which may cause serious and potentially lethal infections (Myc. Tub. Cox. Burnetti)

Safety Requirements:
1. Primary Barriers: Aerosol-tight chamber for work
2. Secondary Barriers: Controlled access to the laboratory

Biosafety Level 4 (high risk)
- Laboratory appropriate for research

- Work is done with dangerous and exotic agents that poses a high individual risk of life-threatening disease, which is transmitted via the aerosol route and for which there is no vaccine or therapy available (Marburg virus, Ebola virus)

Safety Requirements:

1. Primary Barriers: complete full-body air-supplied, positive pressure personal suit (Biosafety Class III cabinets)
2. Secondary Barriers: complete isolated zone in a separate building

A.2 Biohazard Materials

1. Human pathogens (bacteria, fungi, viruses, parasites, prions)
2. All human blood products, tissues and body fluids
3. Cultured cells
4. Toxins
5. Infected tissues

A.3 Recommended Work Practices

1. Practice aseptic culture techniques
2. Keep good record of tissue specimens — source, date etc.
3. Maintain proper containment — where and how to handle specimens in the laboratory

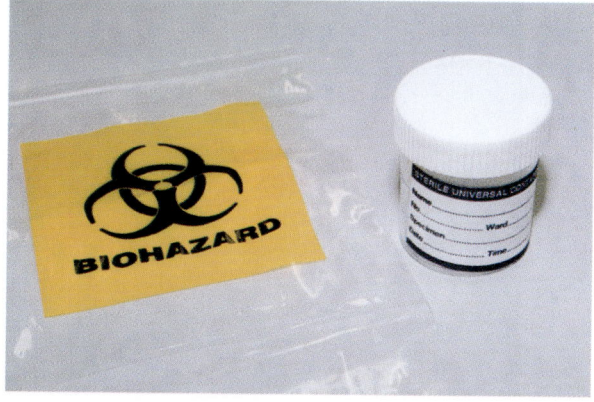

Fig. A.1 Biohazard bag and container.

4. Use appropriate lab wear — gloves (if necessary double gloving), lab coat, proper footwear, safety goggles (if needed) and do not wear protective clothing outside the laboratory
5. Maintain a proper working environment — clean up and disinfect equipment after use! (10% household bleach or Chlorox is an effective disinfectant)
6. Keep laboratory doors closed when experiments are in progress

A.4 General Hygiene

1. Maintain personal hygiene — wash your hands, tie back long hair
2. Do not touch your face/hair with gloves and do not use mobile phones with gloves
3. Do not eat, drink, smoke in the laboratory
4. Avoid talking when doing sterile work

Fig. A.2 Proper disposal of biological wastes.

A.4.1 Pipettes

- No mouth pipetting
- Always use cotton plugged pipettes
- Avoid creating bubbles
- Do not mix fluids in the pipette
- Place contaminated reusable pipettes in a container with disinfectant
- Autoclave contaminated disposable pipettes in an appropriate bag/container before disposal

A.4.2 Syringes and Scalpels

- Use disposable needles
- Never put the cap back on the needle
- Do not use excessive force when fitting a needle or a scalpel
- Dispose in container meant for sharp objects after use

A.4.3 Biosafety Cabinets

- Use at least a class II biosafety cabinet for human cell culture work
- Plan your work in advance

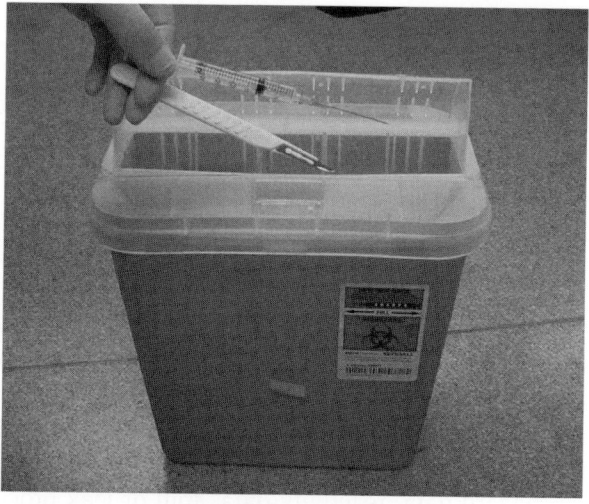

Fig. A.3 Disposal of sharp objects.

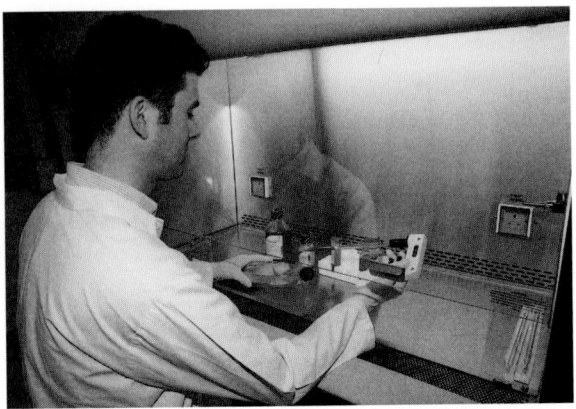

Fig. A.4 Working at the biosafety cabinet.

- Switch on the UV lamp 10 mins before use with front panel closed to maximise sterility
- Switch off the UV lamp before starting work; it is harmful for the eyes and the cultures
- Turn the blower on and leave it running at least 5 mins prior to use
- Wipe bench top with 70% ethanol before use
- Spray items outside the cabinet with 70% ethanol before bringing them in
- Minimise storage of things in the cabinet
- Do not place objects on the front air intake grill
- Clean up spills immediately
- Clean up the cabinet with disinfectant after use
- Close front panel and switch on the UV lamp

A.4.4 Chemicals

A number of chemicals used in the laboratory are hazardous. All manufacturers of hazardous materials are required by law to supply the user with pertinent information on any hazards associated with their chemicals. This information is supplied in the form of Material Safety Data Sheets or MSDS. This information contains the chemical name, CAS#, health hazard data including first aid treatment, physical data, fire and explosion hazard data, reactivity data, spill or leak procedures, and any special precautions needed when handling this chemical.

MSDS information can be accessed on World Wide Web (www.msdssearch.com, www.msds.com). Researchers are strongly urged to make use of this information prior to using a new chemical, and certainly in the case of any accidental exposure or spill. The principal investigator or laboratory head must be notified immediately in the case of an accident involving any potentially hazardous reagents.

The following chemicals are particularly noteworthy:

- Phenol — can cause severe burns
- Acrylamide — potential neurotoxin
- Ethidium bromide — carcinogen

These chemicals are not harmful if used properly: always wear gloves when using potentially hazardous chemicals and never mouth-pipette them. If you accidentally splash any of these chemicals on your skin, rinse the area thoroughly with water immediately, and inform the laboratory safety officer. Discard chemical wastes in appropriate containers.

Important

Do not discard chemical waste down the sink!

A.4.5 Ultraviolet Light

Exposure to ultraviolet (UV) light can cause acute eye irritation. Since the retina cannot detect UV light, you may not realize that you have serious eye damage until 30 mins to 24 hours after exposure. Therefore, always turn off the UV light or wear appropriate eye protection before entering the laboratory or starting to work at the biosafety cabinet.

A.4.6 General Housekeeping

All common areas should be kept free of clutter and all dirty dishes. Since you have only a limited amount of space to call your own, it is to your advantage to keep your own area clean. As you will be using common facilities, all solutions and everything stored in an incubator, refrigerator, etc. must be labelled. In order to limit confusion, each person should use his initials or some other unique designation for labelling plates, etc. Unlabelled material found in the biosafety cabinets, incubators or freezers may be destroyed. Always mark culture vessels with your initials, the date and relevant experimental data, e.g. strain numbers. Each person should be assigned general lab duties that may include keeping track of inventory, making sure a given area is kept clean, or maintaining equipment.

Note

"Think about what you are doing. The best defense is common sense."

A.5 Handling Biological Specimens

Researchers working with primary human tissue samples should know about the existing risks of disease transmission while working with the specimen. The most important and fundamental precaution would be to follow proper work practice in an appropriate laboratory environment designed for primary cell culture work.

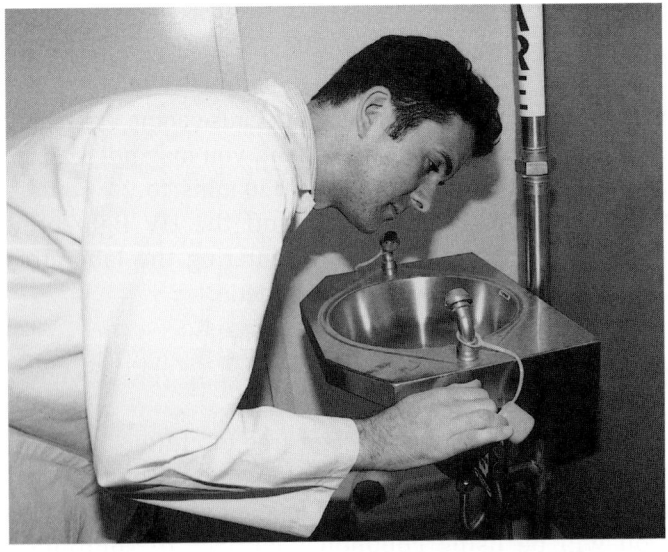

Fig. A.5 Irrigation of the eyes using an eye shower.

In addition, it is mandatory that new researchers joining the group receive proper training in specimen handling and that they are informed about the biological specimens involved. **Remember:** *All primary human tissue samples are potentially contagious!* We also recommend vaccinations against **Hepatitis A/B, Tuberculosis and Tetanus**.

In the event of an "occupational exposure" — contact of potentially infectious material with the body, either via contact with the skin or any mucus membrane, ingestion (swallowing of material) or any other parenteral means (e.g. accidental needle puncture), the following steps should be taken immediately:

1. Stop working
2. Inform your colleagues/lab safety officer immediately
3. Irrigate the skin/mucus/eye area with water
4. Consult a medical doctor immediately
5. Keep a sample of the material for further diagnostic tests

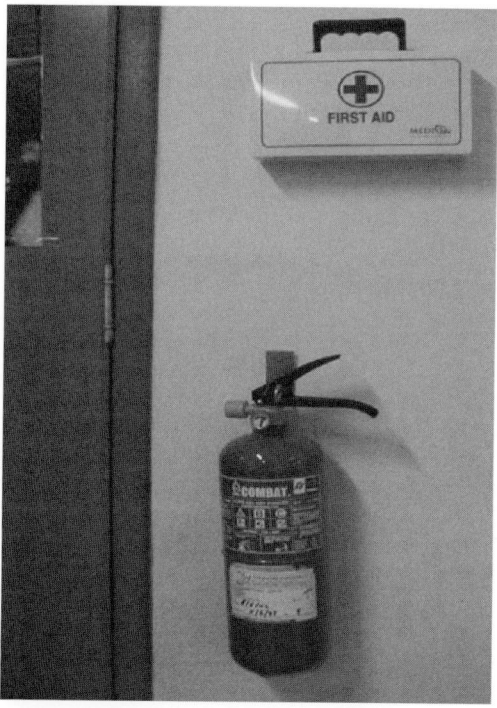

Fig. A.6 Safety kit and fire extinguisher.

A.5.1 Checklist (Update the Equipment Every 2 Months!)

- Equipped first aid kit (disinfectant solution, absorbent gauzes, disposable gloves, adhesive bandage, crepe bandage, antiseptic cream, triangular bandage, plasters, general guidance booklet)
- Working eye shower
- Working fire extinguisher
- Accessible emergency exits
- Available emergency telephone numbers
- Lab staff trained in first aid procedures
- Biosafety manual

B

Introduction to the Cell Culture Laboratory

The Cell Culture Laboratory

This section gives a brief overview about the basic equipment in a cell culture laboratory. Thus it provides a useful check-list for beginners in cell culture work, or researchers planning to set up a cell culture laboratory facility. However these requirements will vary, depending largely upon the role planned for cell culture in the individual research programs. Space and instrumentation requirements may increase when a very active role is anticipated. Sharing of equipment is generally adopted, therefore one must take certain precautions that cross-contaminations are avoided. In line with that, regular maintenance and checking of laboratory instruments and equipment is vital. Taking care and proper handling are particularly important to maintain a safe and clean working environment, and to obtain good results.

B.1 The Laminar Flow Workbench — Biosafety Cabinet

The Biosafety cabinet is the primary workspace to handle human tissue samples (see *Chapter A: Biosafety and Biohazards*). It should protect the user and the sample handled in the workspace. All cell culture work should be carried out in a class II cabinet. The air circulates vertically and is purified through HEPA (high efficiency particulate air) filters that remove 99% of the aerosols and particles.

Together with the cell culture incubator, the biosafety cabinet represents the core of a culture lab and it is important that a suitable place is selected. Areas close to the air-conditioning, windows and doors should be avoided as the air turbulences can interfere with the operation of the cabinet. Environmental monitoring within the safety cabinet can be performed with Soya Broth agar plates. The cabinet works well when there is no growth of bacteria or fungi on these plates.

B.2 The Centrifuge

The centrifuge (refrigerated or non-refrigerated) represents an essential equipment in the lab. It is mainly used in isolating cells

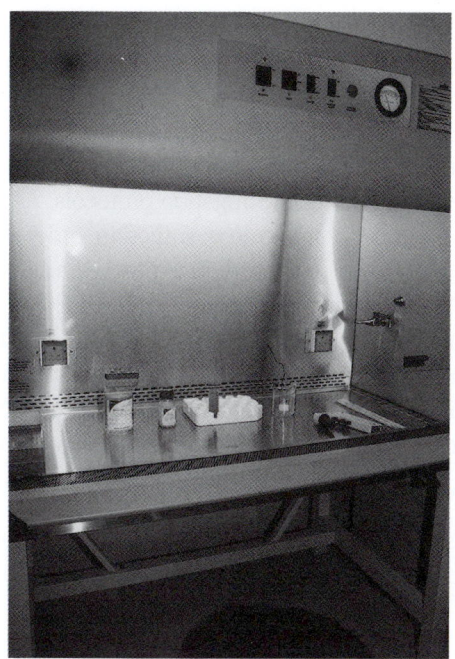

Fig. B.1 A Class II laminar flow biosafety cabinet.

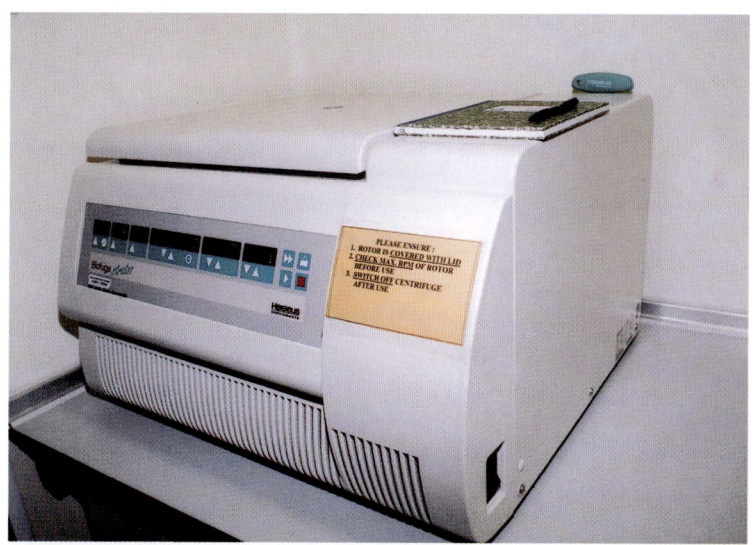

Fig. B.2 A refrigerated centrifuge fitted with a swing bucket rotor.

during density gradient centrifugation or when concentrating cells in suspension. A swing bucket rotor is generally preferred to a fixed angle rotor.

It is however useful if both are available as the swing bucket allows only maximum speed of 6000–8000 rpm, which is sometimes too low for gradient centrifugation protocols. Acceleration and brake forces should be set to low values in order to reduce cell damage and so as not to disturb the pellet/gradient. In general, cells can be safely spun down at a force of 200 g for 10 mins, however, some cells require different protocols (see the following sections). The centrifuge nomograph below assists in converting rpm (revolutions per minute) into rcf (relative centrifugal force, or g-force). Alternatively, the following conversion formula can be used,

$$\text{Relative centrifugal force} = (1.12 \times 10^{-5}) \times \text{rpm}^2 \times \text{radius (cm)}$$

where radius is the furthest distance between the center of the centrifuge spindle to the base of the centrifuge tube.

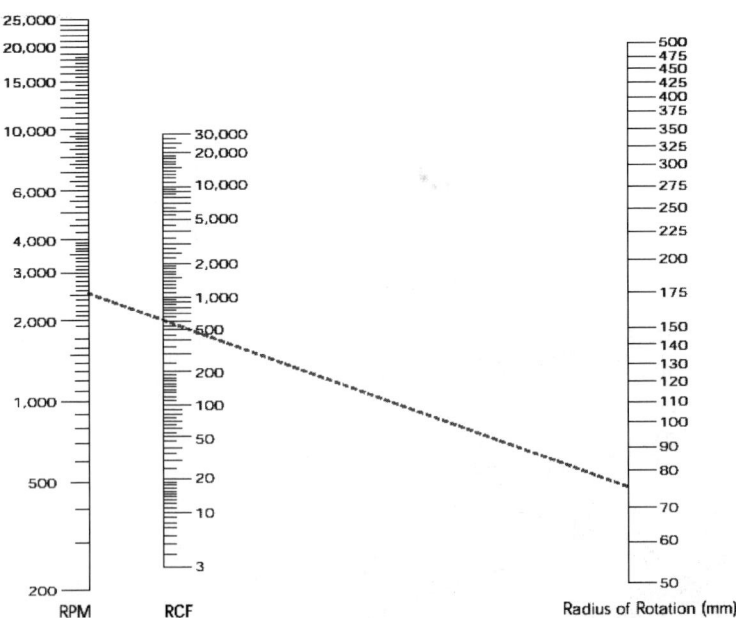

Fig. B.3 A centrifuge nomograph for converting between rpm and rcf. A line is drawn across known radius and value of either rpm or rcf to obtain corresponding unknown.

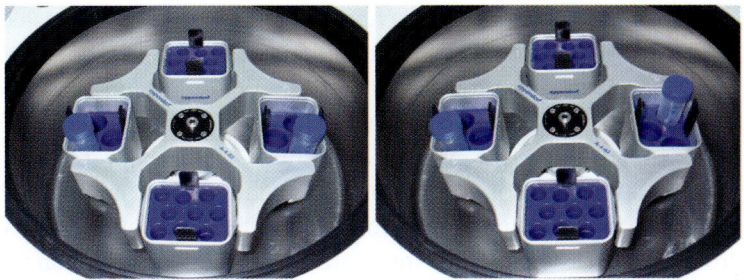

Fig. B.4 Balancing tubes in a centrifuge. *Left*: Tubes wrongly positioned directly opposite each other. *Right*: Tubes correctly balanced diagonally opposite each other.

Always remember to balance the centrifuge correctly, otherwise the rotor can be damaged!

B.3 The Waterbath

The waterbath provides an essential tool in bringing liquids like culture medium or PBS to the right temperature prior to use. It further assists in thawing frozen liquids or tissue specimens. When placing bottles and tubes into the bath, it is important to fix them

Fig. B.5 A waterbath for warming up media and reagents.

and prevent them from getting completely submerged in the water as this increases the risk of contamination through the treads of the bottles. It is a good habit to parafilm all bottles and to spray all items with 70% ethanol when removing from the waterbath.

Similar to the sink/wet space, the waterbath presents a frequent source of contamination since the temperature and the water provide an almost ideal source for bacteria and fungi to grow. It is therefore almost imperative to use additives (such as SystemClean® or Sigmaclean®) to help keep the water clean, as well as changing the water regularly (weekly) and wiping the whole waterbath with 70% ethanol. Distilled water should always be used in the waterbath to prevent corrosion.

B.4 The Sink/Wet space

Although not as essential as the other equipment included in this section, it represents a support facility and should be considered in the space requirements when a laboratory facility is planned. It should be equipped with a draining area where beakers and glassware can be kept for drying. It is strongly recommended that glassware used for cell culture is exclusively used for this purpose and not mixed with glassware for other (e.g. chemical) procedures. Basic washing agents such as hand soap and detergent should be available, as well as disinfectant such as Chlorox.

It has to be pointed out that no toxic chemicals or any biological fluids, including untreated cell culture media, should be poured into the sink. It not only creates a source of contamination, but is also a bio-chemical hazard to the environment!

B.5 The Autoclave

An autoclave plays a central role within the cleaning and sterilisation facility of a cell culture laboratory. It should be in close proximity to the sink and away from aseptic areas like the biosafety cabinet and the incubators. Although the use of disposable plastic ware has led to a decrease in its use, the autoclave is still critical in a laboratory working with tissue

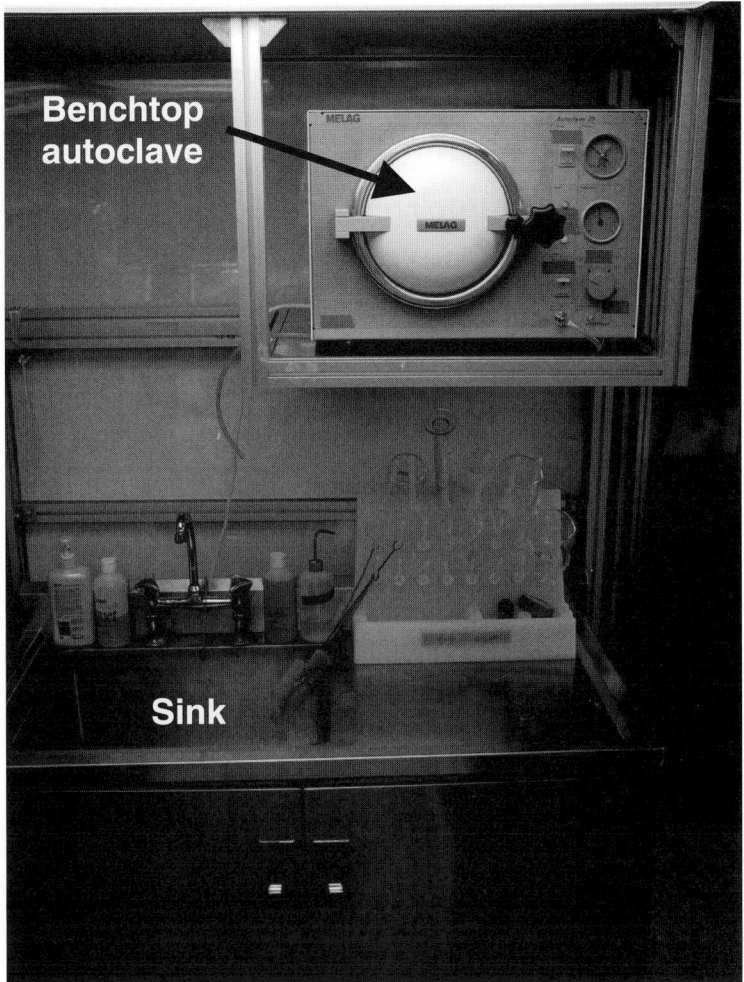

Fig. B.6 An autoclave housed above a sink.

specimens and primary cells as glassware and surgical equipment frequently need to be sterilised.

All items for autoclaving should be properly packed (sterilising bags, cotton drapes, aluminium foil or specific containers) and labelled with a heat sensitive indicator tape to monitor the effectiveness of the procedure. After autoclaving the hot materials should be removed and kept in a place with sufficient ventilation for drying.

Important

Keep temperature charts to avoid heat derived damage to materials.

B.6 The Refrigerator/Freezer

The refrigerator and freezer are important equipment in the cell culture laboratory since essentially all liquid reagents and specimens have to be stored below room temperature. A refrigerator that holds a temperature of ~4°C is required for the storage of media and biological samples (use separate compartments to avoid cross contamination). Culture media are usually light sensitive, thus a refrigerator with only a small front window panel is advisable. A temperature of −20°C is necessary

Fig. B.7 A refrigerator for storing culture media.

for the short-term storage of biochemicals and serum. An ultra-low freezer which goes down to −150°C or a liquid nitrogen tank with −196°C allows long-term storage of cryopreserved cells.

B.7 The Incubator

The incubator is where cell cultures spend the most time, and therefore plays a vital role in maintaining an optimal environment

Fig. B.8 An air-jacketed, self-sterilisable incubator.

for cells to proliferate. Incubators range from simple temperature-controlled boxes to elaborate self-sterilisable, air-jacketed units with temperature, CO_2 and humidity controls. They should be placed as near as possible to the aseptic work area (laminar flow hoods) for the convenience of the cell culturist, as well as to minimise risks of contamination during transportation. Regular cleaning of the incubator interior with 70% ethanol or any anti-microbial agents is necessary to maintain as clean an environment within the incubator as possible. If the incubator has the self-sterilising function, it should be carried out every 1–2 months. Some incubator models have a copper-coated surface inside, which helps to reduce the risk of microbial contamination. Finally, it is important to ensure that incubators have a backup power source in case of unanticipated power failures.

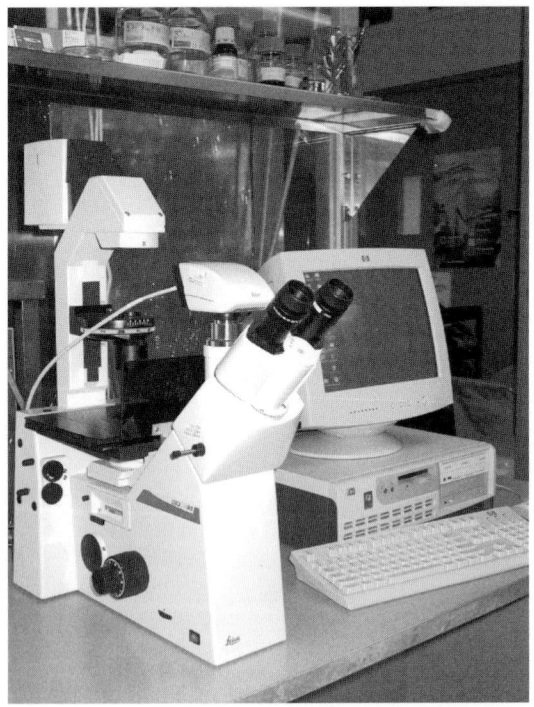

Fig. B.9 An inverted phase contrast light microscope, fitted with CCTV for digital image capture.

Important

> Do not frequently open and close incubator doors — this leads to a drop in temperature, humidity and CO_2 levels, which will affect cell growth.

B.8 The Microscope

A microscope is essential in a cell culture laboratory because it offers the most direct and convenient way of observing cells via visual inspection. Observing one's own cultures using the microscope should become second nature to all cell culturists and it should be done on a regular basis, if not daily. Important information such as cell morphology, cell proliferation rate and culture sterility can be obtained, allowing immediate decisions to be made and any rectification steps promptly taken. Generally, an inverted microscope, equipped with phase-contrast optics and essential filters, would be enough for observing cultures in various culture vessels, as well as histological slides. Ultraviolet optics can also be added for fluorescence observations. Typical objectives fitted on a microscope include ×2.5, ×4, ×10, ×20 and ×40. The final magnification used would be the multiplication of the objective and eyepiece magnifications. When the microscope is not in use, it should be covered to avoid dust settling on the optic systems as well as to prevent damage to eyepiece and objectives.

C

Protocols for Human Cell Culture

All tissues derive from three distinct primary germ layers during embryonic development. The layers are named **Mesoderm**, **Ectoderm** and **Endoderm**. It is essential when culturing primary mammalian/human cells to know where a particular cell type originates from as it gives information about cell characteristics and physiologic behaviour. Cells originating from the Mesoderm include the mesenchymal lineages (osteogenic, chondrogenic, adipogenic and myogenic cells), the urogenital cells of kidney, the circulatory system, the reproductive system and the connective tissue of the skin. Cells from the Ectoderm include the lining cells of the skin (Keratinocytes), the dental tissue (tooth) and neural tissue. Cells from the Endoderm include cells from the gut, the liver (hepatocytes), the pancreas as well as the respiratory tract. We decided to modify this classification slightly by grouping the cell types in this manual into: Mesenchymal cells (derived from the Mesoderm); Epithelial cells (Ecto and Endoderm) including parenchymal cells of the liver; and Bone marrow derived cells (Mesoderm), which include Bone marrow derived stomal cells (mesenchymal cells) and peripheral blood mononuclear cells (haematopoietic cells).

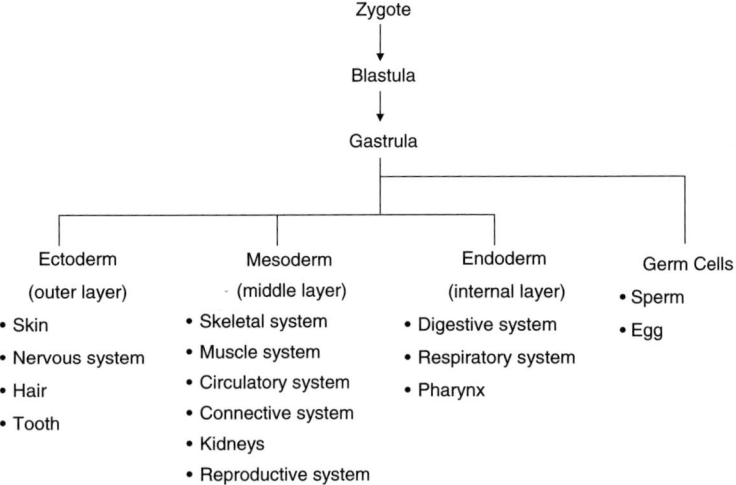

Fig. C.1 Lineage tree. The development of a multicellular organism starts from a single cell, which consecutively divides and gives rise to the 3 distinct germ layers (Ecto, Meso and Endoderm) as well as the germ cells (Sperm and Egg).

When a primary culture from a tissue sample is prepared, it is often heterogeneous in nature. For example, in a keratinocyte culture you will always find some fibroblasts, endothelial cells or melanocytes (see Fig. C.9a). Such a mixture is generally not a problem and could even be beneficial because this closely resembles cells in their natural environment. However, fibroblasts in particular could become detrimental dominant cells. Methods to control this are mentioned in Section C.2.3 Keratinocytes.

Ethical and Legal Requirement

It is important to note that an informed consent from the patient is required before any tissue specimens can be used for research or diagnostic purposes.

C.1 Cell Isolation

To obtain a primary, culture cells must be isolated from a donor tissue or organ. Those cell populations obtained from a tissue specimen usually comprise numerous different cell types. To characterize a specific type of cell we have to isolate a single cell or fractionate the population into the different cell types. There are various methods to achieve this. Easy to perform, but unspecific, are methods like dilution series, where the highest dilution statistically contains only one cell per volume unit or the fractionation of cells depending on their physical properties like gravity sedimentation, centrifugation and sieving. Slightly more specific are methods that separate cells depending on their physiological properties like plating high diluted cell suspension under certain selective pressures (antibiotics, incomplete media).

Biological molecules like oligonucleotides, lectins, antibodies can be tagged with fluorescence dyes or beads and magnetic beads. These molecules bind specifically to their extra- or intracellular ligands (DNA-, RNA-sequences, membrane-proteins, intracellular proteins) and mark the cells which express these ligands. Thus, the application of tagged molecules allows very specific cell selection. The separation itself varies dependent on the used tag. If fluorescence marker were used the separation is performed with a **F**luorescence **A**ctivated **C**ell **S**orter (FACS). For this purpose each cell gets entrapped in a droplet of charged fluid. These droplets pass the optical unit of the FACS one by one to measure the fluorescence activity of each single cell. The fluorescence detector controls an electromagnetic field downstream of the optical unit which deflects the charged droplets dependent on their fluorescence signal into a positive or negative fraction. Cells marked with magnetic beads can be separated with a **M**agnetic **A**ctivated **C**ell **S**orter (MACS). Here the cells are led by a strong magnetic field. The negative cells pass the field unhindered whereas the positive cells are hold back within the magnetic field. After removal of the magnetic field the positive cells can be flushed out and harvested. Properly applied FACS and MACS can be very powerful tools to separate cells even from highly heterogeneous cell populations. Another method is **L**aser **C**apture **M**icrodissection (LCM). Here a laser beam focally activates a special transfer film which bonds specifically to cells identified and targeted by microscopy within the tissue section. The transfer film with the bonded cells is then lifted off the thin tissue section, leaving all unwanted cells behind (which would contaminate the molecular purity of subsequent analysis or culture).

Note

- It is almost impossible to separate cell populations completely. Unspecific bindings, cross-reaction or suboptimal markers cause contaminations of the target fraction with unwanted cell types.
- Selection with specific markers must be seen in context, depending on the constantly changing expression pattern of a cell. Some markers represent certain states of a cell and are expressed periodically. Thus, the same cell can be positive and negative for one marker, depending on the time of analysis.

When primary cell cultures are subcultured they are referred as "cell strains".

C.2 Mesenchymal Cells

The cells from the mesoderm form connective tissue like bone, cartilage, fat and muscles, as well as the endothelium. The cellular morphology is more or less spindle-shaped (fibroblast-like) and the cells are anchorage dependant. They are usually quite robust and are relatively easy to culture. In a co-culture system, they tend to be the dominant cell type. Primary mesenchymal cells can be subcultured multiple passages (note: cell dedifferentiation occurs!) before senescence sets in. Mesenchymal cells can be isolated via enzymatic or explant methods.

C.2.1 Adipose Cells

A culture of adipocytes can be prepared from an adipose tissue biopsy or from liposuction aspirates. These cultures are frequently "contaminated" with vascular endothelial cells.

Enzymatic Digestion

- Wash adipose tissue with PBS
- Shred adipose tissue with sterile scissors to obtain tissue pieces that are as small as possible, and vortex for further disaggregating
- Repeat 3 times with fresh PBS
- Add 0.075% collagenase A and stir using magnetic stirrer for 30 mins at 37°C in a water bath
- Centrifuge at 1200 g for 10 mins and decant supernatant

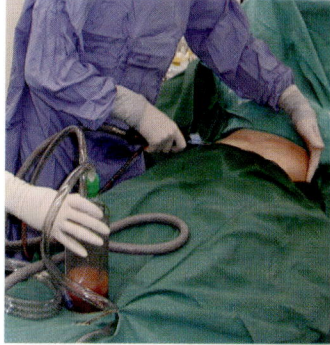 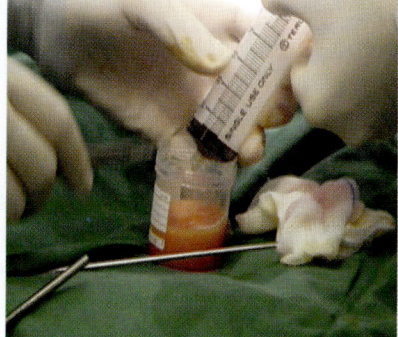

Fig. C.2a Harvesting abdominal adipose tissue from the liposuction aspiration canula.

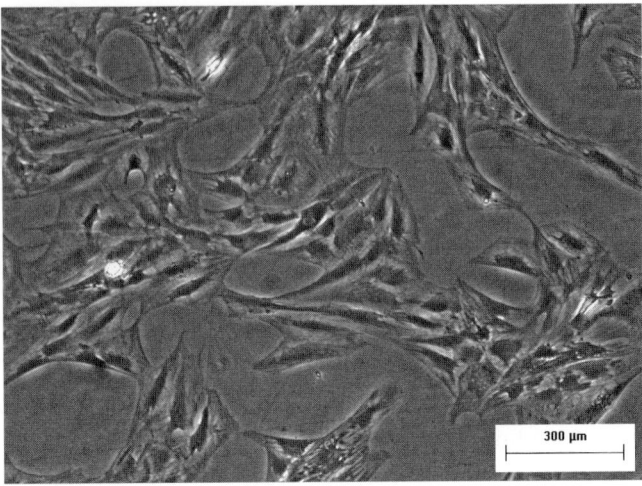

Fig. C.2b Adipose cells in culture ×100.

- Add 10 ml PBS followed by 160 mM (final concentration) NH$_4$Cl to the pellet, and incubate for 10 mins at room temperature to lyse red blood cells
- Centrifuge at 1200 g for 10 mins and decant supernatant
- Resuspend pellet with a micropipette using 1 ml of complete DMEM
- Plate in 75 cm^2 flasks with 9 ml media

Requirements

1. PBS
2. 0.075% Collagenase A
3. Media: DMEM (high glucose) supplemented with 10% FBS and 1% Penincillin–Streptomycin
4. 160 mM NH$_4$Cl

C.2.2 Chondrocytes

In the body, there are three different types of cartilage: elastic (ear, nose), fibrous (rib, intervertebral disc) and hyaline (articular). The majority of the cartilage consists of extracellular matrix (ECM) built up from collagen (Type II) and glycosaminoglycans (GAGs). Cartilage is avascular and is covered *in vivo* by a thin fibrous membrane called the perichondrium, which supplies the tissue with necessary nutrients. The most reliable and efficient way to obtain a chondrocyte culture from a cartilage biopsy is via enzymatic digestion. Although an explant culture system also works, a much longer time is needed and usually fewer cells are obtained.

Enzymatic Digestion

- Collect the freshly harvested cartilage pieces in PBS solution containing 2% Penicillin–Streptomycin (100 U/ml; 100 μg/ml)
- Rinse them twice in PBS
- Remove remaining perichondrium
- Manually mince the cartilage pieces into chips of 1–2 mm
- Transfer the chips into petri dishes and add collagenase solution (Collagenase II, 2 mg/ml in serum-free culture media)
- Incubate the chips for 16–24 hrs depending on the chip size
- Transfer the digested solution into 50 ml centrifuge tubes and add fresh culture media (containing serum)
- Vortex the solution for 5 mins
- Filter the solution through 70 μm nylon sieves
- Centrifuge the cells at 600 g for 10 mins
- Wash 3 times using complete media
- Resuspend the pellet and transfer for culturing into culture flasks

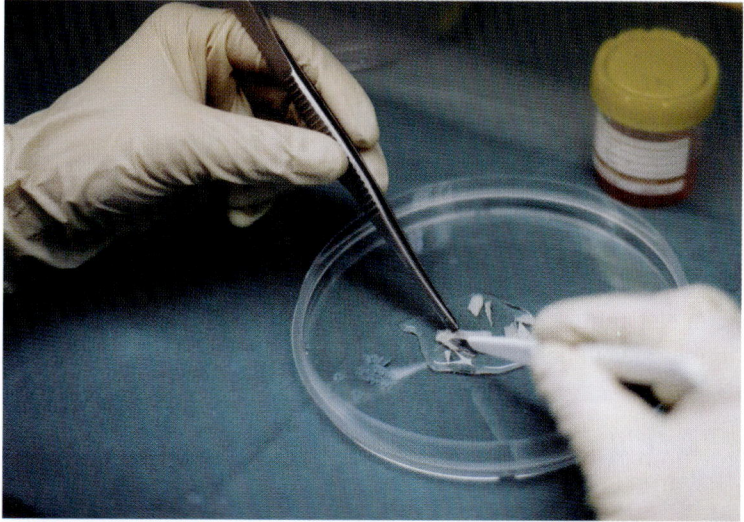

Fig. C.3a Manual mincing of cartilage pieces.

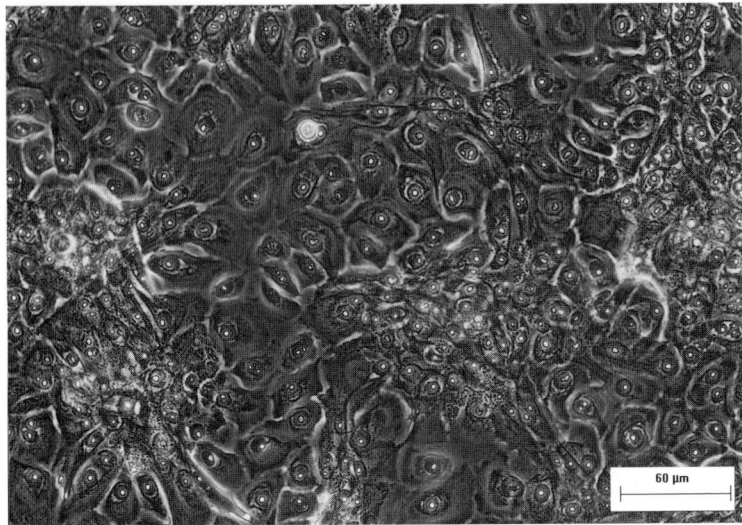

Fig. C.3b Chondrocytes in culture ×200.

Requirements

1. Media: Ham's F–12 (supplements: 10% FBS, 1% Penicillin–Streptomycin, 25 mg Ascorbic acid, 1% 200 mM L–Glutamine)
2. Collagenase: Type II 194 U/mg; working solution 2 mg/ml (sterile filtered)

C.2.3 Endothelial Cells

Endothelial cells have a surface resembling the cobblestone morphology similar to keratinocytes and are derived from the endoderm. Like keratinocytes, they have special requirements to the media and the culture surface, and demand some experience with cell culture. For the isolation of endothelial cells, the human umbilical vein has shown to be a reliable source for cell harvesting.

HUVEC (Human Umbilical Vein Endothelial Cells) Isolation

• Place the fresh umbilical cord sample into sterile Ringer's or Hartmann's solution

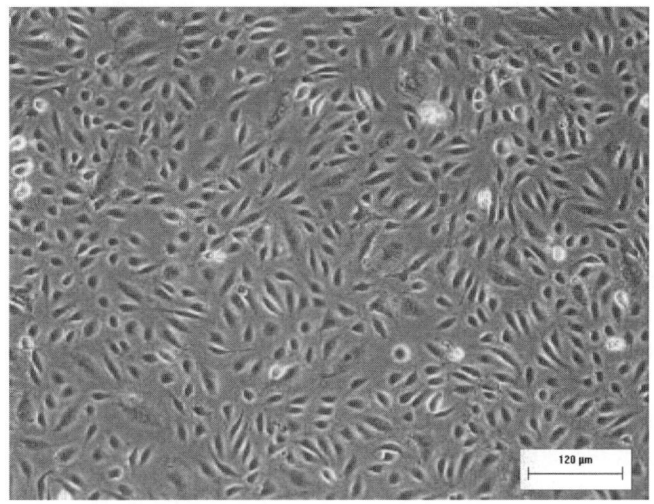

Fig. C.4 HUVEC in culture ×100.

- Clamp a luer lock syringe to one end of the umbilical vein and a luer lock adapter to the other end, and rinse it properly with 37°C PBS to wash out the remaining blood
- Fill the vein with warm dispase solution or collagenase solution (clamp both sides), and incubate it for 40 mins in the incubator
- Open the lock and release the dispase cell suspension from the lumen — flush it twice with warm PBS
- Centrifuge the cells at 200 g for 10 mins
- Discard the supernatant and resuspend the endothelial cell pellet in **H**uman **E**ndothelial **G**rowth **M**edia (HEGM)
- Place about 2 ml of endothelial cell suspension per 75 cm^2 gelatine precoated culture flasks
- Change media every 2–3 days

Requirements

1. Media: HEGM; supplemented with 5% FBS; 1% Penicillin–Streptomycin; 1% Fungizone and 2 mM L–Glutamine. Alternatively M199 can be used
2. Gelatine Coating: 0.1% gelatine (sterile filtered) solution

3. Collagenase: Type II; 0.5% Collagenase diluted into 0.1% BSA in PBS (sterile filtered)
4. Dispase: 2.5 U/ml in PBS (sterile filtered)

C.2.4 Fibroblasts

Fibroblasts are generally regarded as the easiest mammalian cells to culture. They are relatively undifferentiated; meaning that the use of special reagents to induce differentiation is unnecessary. In addition, fibroblasts are robust and will attach and proliferate on a wide variety of culture vessels and biomaterials. In this section, the isolation and maintenance of fibroblasts derived from human dermis is described.

Explant Method

- Separate epidermis and dermis either by mechanical means using sterile forceps and scissors, or by enzymatic digestion (see *Separating the Epidermis from Dermis*, pg. 55)
- Cut dermis into pieces not bigger than strips of 2–3 mm

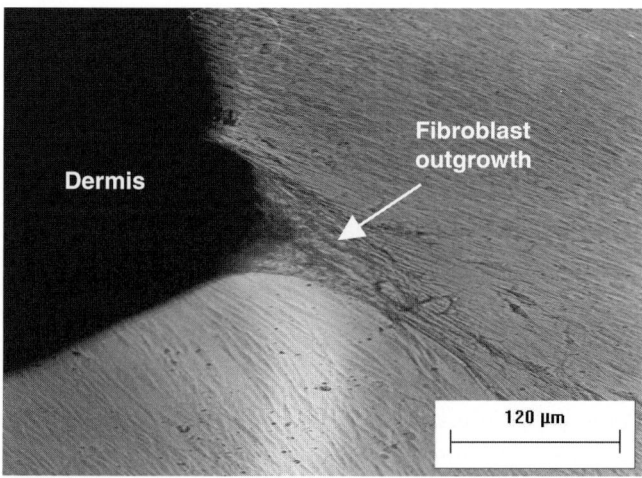

Fig. C.5a Fibroblast outgrowth from explant culture ×100.

- Transfer samples into culture flasks or petri dishes containing just enough medium [DMEM supplemented with 1% Penicillin–Streptomycin (50 U/ml; 50 μg/ml) and 10% FBS] to wet the entire vessel surface so as not to disturb the tissue
- Distribute samples over the culture surface to maximise space
- Leave the culture undisturbed in the incubator for 3–5 days before the first media change

Enzymatic Digestion

- Separate epidermis and dermis either by mechanical means using sterile forceps and scissors or by enzymatic digestion (see *Separating the Epidermis from Dermis*, pg. 55)
- Mince dermal samples manually into small pieces and transfer into a petri dish or well plate containing 0.2 mg/ml collagenase type I in serum-free DMEM. 0.1% trypsin can also be added
- Incubate overnight at 37°C
- Filter supernatant through 100 μm nylon sieves to remove tissue debris
- Centrifuge at 200 g for 10 mins, and plate at desired density (typically 1–4000/cm² for routine culture) in culture flasks with DMEM supplemented with 1% Penicillin–Streptomycin and 10% FBS

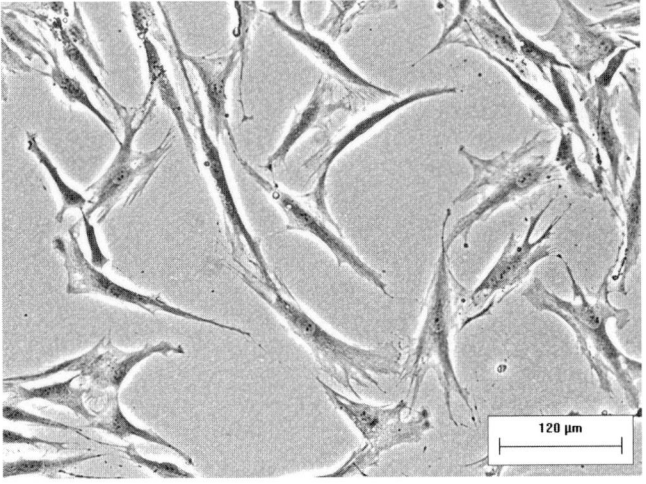

120 μm

Fig. C.5b Isolated fibroblasts in culture ×100.

Subculturing

- At 80–90% confluence, aspirate culture medium and add 0.25% trypsin-EDTA to culture flasks. (Typically, the volume of trypsin used is 5 ml for a 75 cm^2 flask)
- Incubate for 5 mins at 37°C. (Incubation time may be increased if needed depending on confluence, age of culture etc.)
- Check under the microscope to ensure that all cells have detached
- Tap lightly on the sides of the flasks to dislodge detaching cells
- Add an equal amount of complete DMEM or trypsin inhibitor to stop trypsin activity and collect the cell suspension in a 15 or 50 ml tube
- Centrifuge cell suspension at 200 g for 10 mins and discard supernatant
- Resuspend in complete DMEM, count if desired, and split into new culture flasks in 1:2 or 1:3 ratio

Requirements

1. Media: DMEM (high glucose) supplemented with 10% FBS and 1% Penicillin–Streptomycin
2. 0.25% Trypsin–EDTA
3. 2 mg/ml Collagenase type I (optional, depending on isolation method; sterile filtered)
4. Trypsin inhibitor (optional)

C.2.5 Myocytes

Skeletal Muscle Cells

Skeletal muscle cells are derived from myogenic satellite cells, which lie between the basal lamina and the sarcolemma of differentiated muscle fibres. Satellite cells isolated and cultured *in vitro* are known as myoblasts. Myoblasts proliferate and migrate at random in culture, and then align themselves and differentiate to form multinucleated myotubes after about 10 days in culture. Once differentiated, subculture is no longer possible. Typically, monolayer myoblast cultures can be passaged 3–4 times via trypsination.

Enzymatic digestion

- Trim off non-muscle tissue from biopsy and rinse in PBS
- Cut tissue into fragments parallel to the fibres and wash in PBS prior to weighing
- Cut fragments further into 1 mm^3 pieces and add 0.15% Pronase solution in serum-free Ham's F–12 or PBS for 1 hr at 37°C. (Use 15 ml pronase solution for 1–3 mg tissue)
- Shake gently at 15 min intervals
- Vortex the tissue-cell suspension mixture and filter through a 100 μm nylon sieves
- Centrifuge collected cell-suspension at 200 g for 10 mins
- Resuspend cell pellet gently in Ham's F–12 medium supplemented with 20% FBS and 1% Penicillin–Streptomycin (50 U/ml; 50 μg/ml)
- Count cells with a haemocytometer and plate at a density of 2000/cm^2
- Change medium very gently after 24 hrs and subsequently every 3–4 days
- To maximise isolation efficiency, re-digest left-over tissue

Subculturing

- At 70–80% confluence, aspirate culture medium and add 0.25% trypsin-EDTA to culture flasks. (Typically, the volume of trypsin used is 5 ml for a 75 cm^2 flask)
- Add just enough 0.25% trypsin-EDTA to cover the entire culture surface, incubate 5 mins at 37°C. (Leave longer if necessary)
- Check under microscope to ensure that all cells have detached
- Tap lightly on the sides of the flasks to dislodge detaching cells
- Add at least an equal amount of complete Ham's F–12 or trypsin inhibitor to neutralise trypsin and collect the cell suspension in a 15 or 50 ml tube
- Centrifuge cell suspension at 200 g for 10 mins and discard supernatant
- Resuspend in complete media, count if desired, and split into new culture flasks in desired ratio

Requirements

1. PBS
2. Media: Ham's F–12 supplemented with 20% FBS and 1% Penicillin–Streptomycin 20 mM HEPES
3. 1.5 mg/ml Pronase in serum-free Ham's F–12 or PBS, with 1% Penicillin–Streptomycin (sterile filtered)
4. 0.25% Trypsin-EDTA

Smooth Muscle Cells

This section describes the isolation of smooth muscle cells (SMCs) from thoracic aorta samples of heart transplant donors.

Enzymatic digestion

- Rinse aorta sample extensively in HBSS
- Cut open aorta along the longitudinal axis and remove the inner endothelium by scraping the cell layer off using a scalpel
- Peel off the intima, if clearly defined, using forceps
- Harvest the media (Avascular middle layer of the artery wall, composing alternate layers of elastic fibres and smooth muscle cells) by peeling strips of it from the outer adventitia
- Remove as much of the media as possible and rinse the strips in HBSS
- Cut media strips into small pieces of 1–2 mm
- Rinse the tissue in HBSS again and weigh the tissue in a sterile container — the weight of the tissue will determine how much enzyme solution to use [tissue (g) to enzyme (ml) ratio is 1:5 w/v]
- Rinse the tissue with a small volume of 3 mg/ml collagenase type I solution in serum-free M199, aspirate and add fresh collagenase solution in the correct ratio
- Incubate for 30 mins at 37°C, preferably with gentle shaking on a shaker
- Prepare 1 mg/ml elastase solution in serum-free M199, of the same volume as collagenase solution used
- Add the elastase solution into the tissue/collagenase mixture and continue to incubate at 37°C on a shaker, checking every 30 mins until all the tissue has been digested

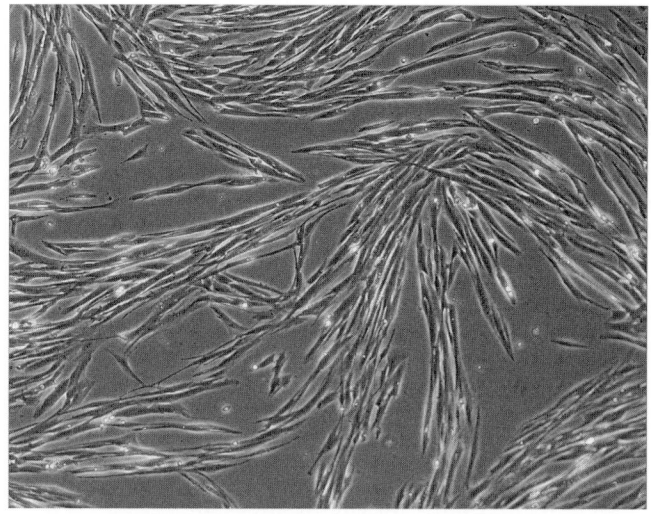

Fig. C.6 Smooth muscle cells in culture ×100.

- Filter the digested tissue/enzyme mixture through a 100 μm nylon sieves to remove large debris
- Centrifuge collected cell-suspension at 200 g for 10 mins and discard supernatant
- Resuspend cell pellet gently in high-glucose DMEM supplemented with 20% FBS and 1% Penicillin–Streptomycin (50 U/ml; 50 μg/ml)
- Count cells with a haemocytometer and plate at a density of 15,000/cm^2 (seeding efficiency typically <20%)
- Change medium very gently after 2 days and subsequently every 2 days

Subculturing

- At 70–80% confluence, aspirate culture medium and add 0.25% trypsin-EDTA to culture flasks. (Typically, the volume of trypsin used is 5 ml for a 75 cm^2 flask)
- Add just enough 0.25% trypsin-EDTA to cover the entire culture surface, incubate for 5 mins at 37°C. (Leave longer if necessary)
- After incubation, check under microscope to ensure that all cells have detached
- Tap lightly on the sides of the flasks to dislodge detaching cells

- Add at least an equal amount of complete DMEM or trypsin inhibitor to neutralise trypsin and collect the cell suspension in a 15 or 50 ml tube
- Centrifuge cell suspension at 200 g for 10 mins and discard supernatant
- Resuspend in complete media, count, and re-plate into new culture flasks at a density of 15,000/cm^2

Requirements

1. HBSS
2. Media: DMEM (high glucose) supplemented with 20% FBS and 1% Penicillin–Streptomycin.
3. 3 mg/ml collagenase type I in serum-free M199, with 1% Penicillin–Streptomycin (sterile filtered).
4. 1 mg/ml elastase in serum-free M199, with 1% Penicillin–Streptomycin, adjusted to pH ~6.8 with 1M HCl (sterile filtered).
5. 0.25% Trypsin-EDTA

C.2.6 Osteoblasts

There are four basic bone types in the human skeleton, which have descriptive names based on their shape: long, short, flat, and irregular.

The macroarchitecture of bone consists of an outer fibrous sheet, the periosteum, sitting on the cortical layer of the bone. The cortical layer gives way to a more spongeous structure — the trabecular or spongeous bone in the centre, and finally the bone marrow in the centre of the bone.

Bone tissue contains two distinct cell types: the osteoblast or bone-forming cell; and the osteoclast or bone-resorbing cell. The structural and functional characteristics of bone as well as bone metabolism are characterised by the complex interplay of these cells under the influence of local and systemic factors. Under physiologic conditions, bone undergoes a constant remodelling with resorption and renewal resulting in tissue homeostasis. During osteogenesis, mesenchymal cells differentiate along the osteogenic pathway into osteoblasts and osteocytes. The cellular morphology of osteoblast is spindle-shaped. The terminally differentiated osteocyte does not divide and has its typical polygonal shaped cytoplasmic extrusions in multiple directions. The cells have a well developed golgi apparatus and endoplasmic reticulum necessary for secretion of proteins.

Periosteum is characterised by two distinct layers: the outer fibrous layer and the inner cambial layer facing towards the cortex of the bone. The outer sheet consists predominantly of fibroblast whereas the inner cambial layer harbours the vasculature and the bone marrow precursor cells.

This chapter summarises the most commonly used protocols to establish a culture for the different osteoblast-like cells: *osteoblasts* and *periosteal cells*. There are 3 major methods to obtain a culture from a bone/periosteum biopsy.

Wash-out Method

- Mince the sample manually into smaller pieces and transfer into a 50 ml centrifuge tube
- Add culture media and vortex the sample on a vortex shaker for 5 mins
- Take the supernatant and centrifuge at 200 g for 10 mins
- Discard the supernatant and resuspend the pellet in media for culture
- First media change after 5 days

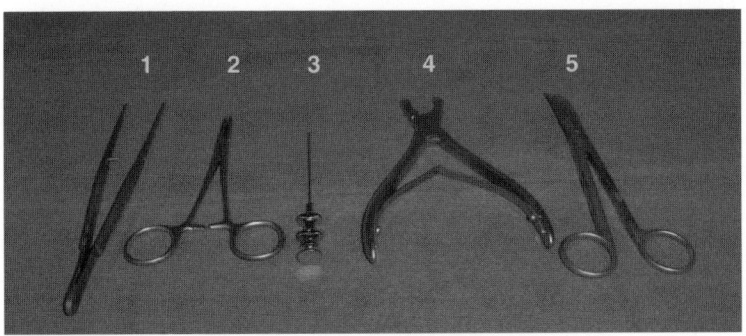

Fig. C.7a Surgical equipment for bone specimens.

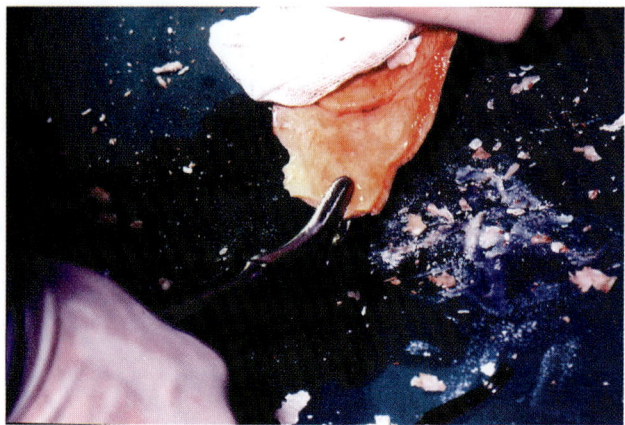

Fig. C.7b Breaking down bone chips from a human calvarial tissue sample.

Equipment Required

1. Strong dissecting forceps with teeth
2. Strong dissecting ligature clamp
3. Bone marrow aspiration needle
4. Bone rongeur
5. Strong dissecting curved scissors

Explant Method

- Mince the samples manually into smaller pieces and transfer into culture flasks
- Add media so that all samples are completely submerged
- First media change after 1 week

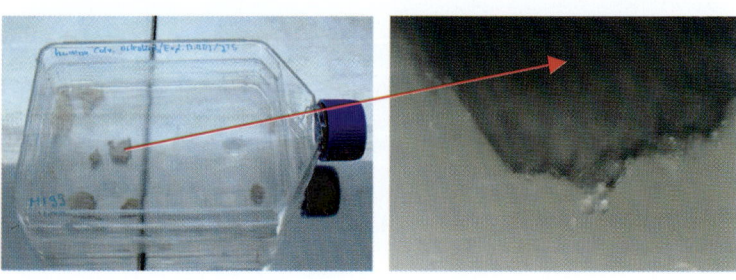

Fig. C.7c Cortical bone explant in culture.   **Fig. C.7d** Osteoblasts migrating out from the bone specimen.

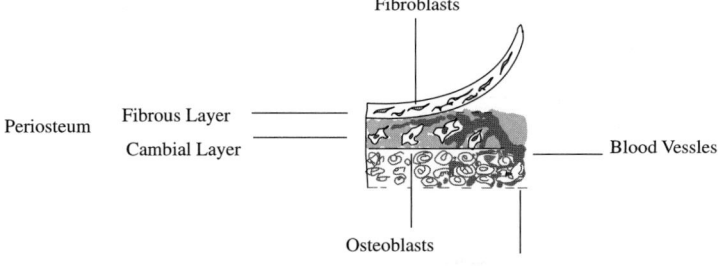

Fig. C.7e The periosteum is located on top of the cortical bone layer. It contains two distinct structures the inner cambial layer (with good vascular supply and predominantly osteoblast-like cells) and the outer fibrous layer (predominantly fibroblasts).

Enzymatic Digestion

- Mince the samples manually into smaller pieces and transfer into a petri dish containing 0.2 mg/ml collagenase A/type I solution (dissolve collagenase A in HBSS or any other Ca^{2+} free buffer)
- Add 0.05% trypsin–EDTA if desired (1:1 ratio)
- Incubate the tissue/collagenase plates for 2 hrs
- Take the supernatant and centrifuge at 200 g for 10 mins
- Resuspend the pellet in complete media
- Wash twice with complete media
- Resuspend the cells and transfer into culture flasks for culturing

Requirements

1. 0.2 mg/ml Collagenase type I solution (optional, depending on isolation method; sterile filtered)
2. Media: M199, DMEM, Mc Coy supplemented with:
   - 10% FBS
   - 1% Penicillin–Streptomycin
   - L–Glutamine
3. Osteogenic differentiation media should be supplemented with:
   - 0.1 μM dexamethasone
   - 100 μM ascorbate–2–phosphate
   - 10 mM β–glycerophosphate
   - 2.5% HEPES

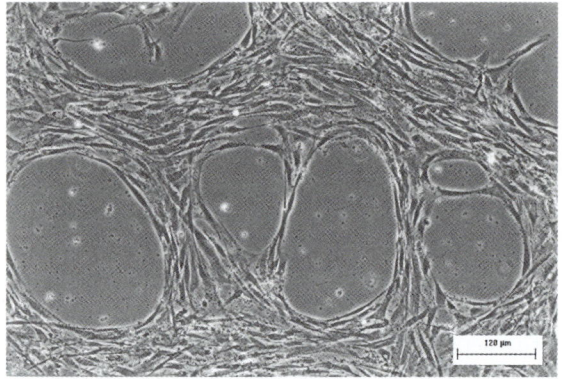

Fig. C.7f Calvarial osteoblasts in culture ×100.

Osteoblasts

Osteoblast-like cells can be isolated from the trabecular (cancellous) or the cortical portion of the bone.
* Collect the bone biopsies in PBS or physiologic saline solution
* Mince the bone manually into small fragments (0.5 × 0.5 mm) using a bone rongeur
* Establish a culture via explant, enzymatic digestion or wash out procedure

Periosteal Cells

* Place the freshly harvested periosteal sheet in PBS to prevent it from drying out

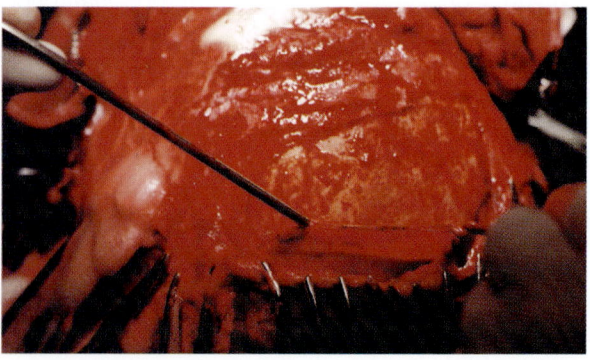

Fig. C.7g Lifting of a periosteal sheet from a human skull.

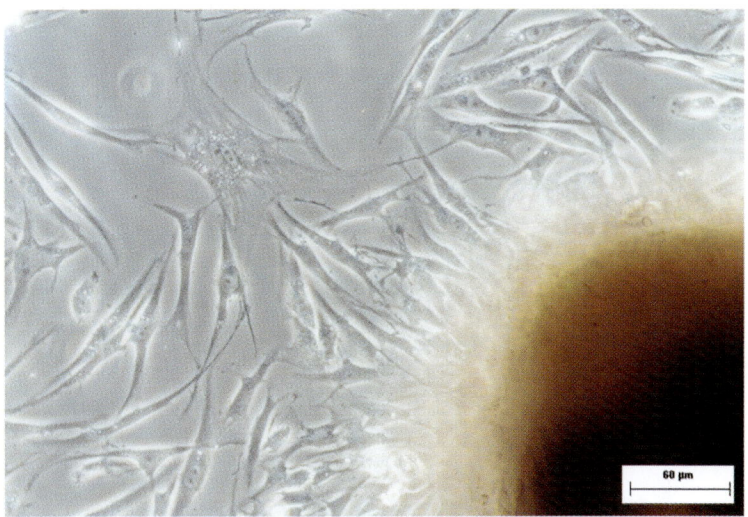

Fig. C.7h Periosteal explant culture ×200.

- Dissect the inner cambial layer (attached to the bone) from the outer fibrous sheet
- Mince the layer manually into smaller pieces
- Establish a culture via explant (add very little media, e.g. 5 ml for a 75 cm^2 flask, to allow the sheet to attach) or via enzymatic digestion

C.2.7 Dental Cells (Periodontal Ligament Fibroblasts & Alveolar Osteoblasts)

The cells of the **Periodontal Ligament** (PDL) maintain and repair alveolar bone and cementum. It is a reservoir from which bone and cementum forming cells are derived; precursor cells are formed from stem cells in the bone marrow, and from there migrate into the PDL. Mature PDL is a highly vascular and cellular tissue. Fibroblasts are the most abundant cell type in the PDL, and are aligned along and between the collagen fibres.

The part of the maxilla and mandible that supports and protects the teeth is known as alveolar bone. Apart from its supportive function, the bone of the maxilla and mandible is also attached to muscles, providing a framework for the bone marrow and acting as a reservoir for ions, in particular calcium.

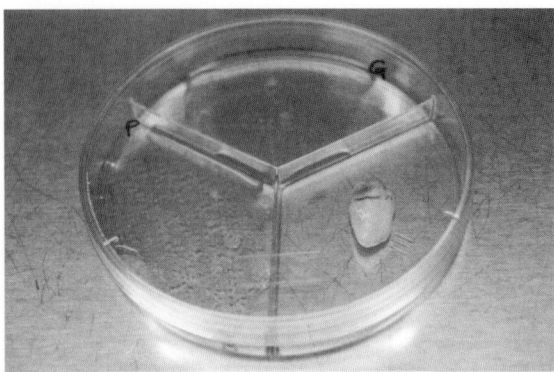

Fig. C.8 Human molar tooth in a chambered petri-dish with freshly isolated Peridontal Ligament Fibroblasts (P) and Gingival cells (G) resuspended in media.

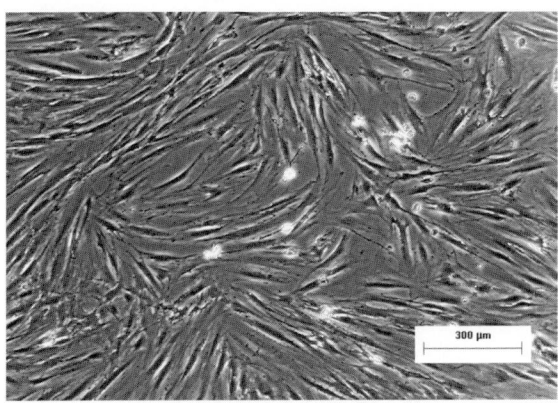

Fig. C.9a PDLF in culture ×40.

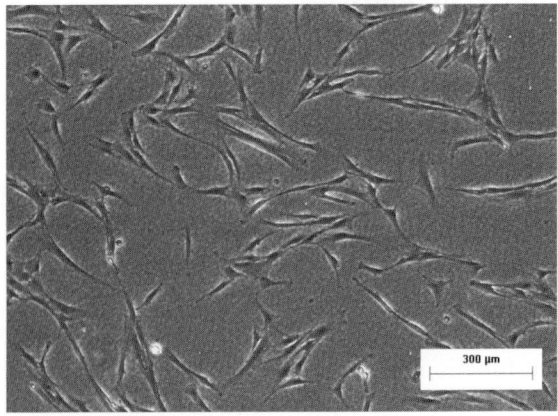

Fig. C.9b Alveolar osteoblasts in culture ×40.

Explant Culture

- Rinse premolar or upper third molar samples in sterile PBS containing 2% Penicillin–Streptomycin solution (100 U/ml and 100 μg/ml)
- Sterilise the samples in biopsy media: DMEM with high glucose (4500 mg/ml), 10% FBS, 2% Penicillin–Streptomycin, 250 μg/ml gentamicin sulphate and 2.5 μg/ml amphotericin B
- Obtain PDL explants by scraping the middle third of the root surface using a sterile surgical blade
- Avoid contamination from gingival or periapical granulation tissues
- Culture PDL explants in DMEM with high glucose (4500 mg/ml), 10% FBS and 2% Penicillin–Streptomycin (100 U/ml and 100 μg/ml)
- Dice alveolar bone explants attached to root surface into small pieces and culture in M199 containing 10% FBS and 2% Penicillin–Streptomycin
- Carry out osteogenic induction with culture media containing 20 μg/ml Ascorbic acid, 10 mM β–Glycerolphosphate, and 10^{-7}M Dexamethasone

Requirements

1. PBS
2. 250 μg/ml gentamicin sulphate
3. 2.5 μg/ml amphotericin B
4. Media — DMEM (high glucose) supplemented with 20% FBS and 2% Penicillin–Streptomycin
   — M199 supplemented with 20% BS and 2% Penicillin–Streptomycin
5. 20 μg/ml Ascorbic acid
6. 10 mM β–Glycerol Phosphate
7. 10^{-7}M Dexamethasone

C.3 Epithelial Cells

Epithelial cells, like airway or skin-lining cells, are more difficult to culture. The cells have a cobble-stone morphology and usually tend to grow in sheets and need direct cell-to-cell contact for growth. Multiple passaging is difficult and sometimes requires feeder layer systems. Epithelial cells can be cultured under serum or serum-free conditions. A commonly observed problem in the culture of epithelium is the overgrowth by stromal cells (e.g. fibroblasts, vascular endothelial cells)

The parenchymal epithelial cells of the liver, pancreas and the gut are responsible for absorption and secretion. Epithelial cells are usually isolated via enzymatic methods.

C.3.1 Respiratory Epithelial Cells

The isolation of primary human respiratory epithelial cells from a piece of nose or tracheal mucosa requires some prior experience in cell culture techniques. Many researchers will find it difficult to get access to sufficient amounts of the tissue. If the culture is established however, it provides an excellent study system for the respiratory tract.

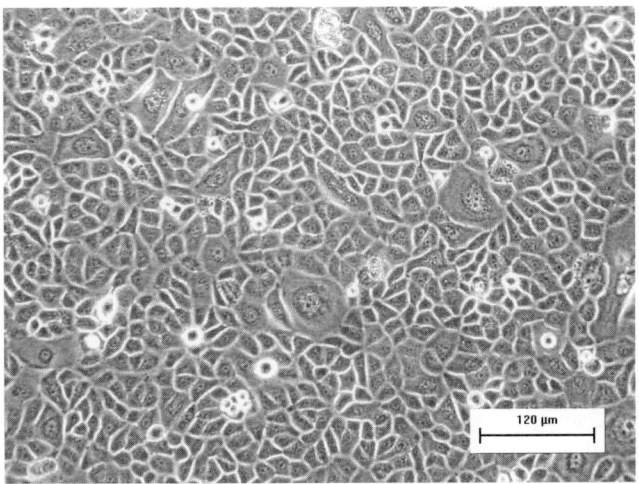

120 µm

Fig. C.10 Respiratory epithelial cells in culture × 100.

Enzymatic Digestion

- Collect the mucosa epithelium in a 50 ml centrifuge tube filled with transport media
- Discard the transport medium and rinse tissue specimen with PBS
- Place the specimen into dispase solution at 4°C for 12 hrs.
- Coat culture vessels with freshly prepared Collagen A before culture. (Use about 2.5 ml of Collagen A for a 25 cm^2 flask; incubate the vessel with the liquid for 30 mins, take out the remaining fluid and gently rinse with PBS)
- Transfer the digested tissue into a petri dish and carefully dissect the yellow shining mucous layer from the underlying fibro-cartilage tissue
- Centrifuge the cell suspension at 300 g for 5 min, discard the supernatant and wash with 10 ml HBBS
- Centrifuge again and resuspend the pellet in approximately 5 ml of complete media
- Take 10 μl of cell suspension and add an equal amount of 1% acetic acid to achieve erythrolysis
- Add 10 μl of the above describe cell solution to 10 μl trypan blue for cell counting and viability assay
- Plate ~3 × 10^5 cells per coated 25 cm^2 flask
- First media change after 2 days

Requirements

1. PBS
2. HBSS
3. Collagen A
4. Dispase II
5. 1% Acetic Acid (in PBS)
6. Enzyme solution (sterile filtered):
   - Dispase 1:10 [1 ml Dispase II (24 U/ml) in 9 ml PBS]
   - Collagen A 1:1 (1 ml Collagen A solution + 1 ml PBS)
   - Use in a final concentration of 1 ml/10 cm^2 culture plate surface

- 1% Acetic Acid (reconstitute from 0.5 ml Acetic Acid (99% ig) + 49.5 ml PBS)
7. Culture media: Bronchial Epithelial Cell Growth Medium (serum free) supplemented with:
   - 10 μg/ml Insulin
   - 5 μg/ml Transferrin
   - 100 ng/ml hydrocortisone
   - 25 ng/ml epidermal growth factor
   - 1% Penicillin–Streptomycin
8. Transport media: RPMI 1640 supplemented with:
   - 10% FBS
   - 1% Penicillin–Streptomycin

C.3.2 Oral Mucosa Cells

The squamous epithelium lines the entire gastro-intestinal tract from the oral cavity to the rectum. The architecture and function of the mucosal lining varies according to the anatomical location. In the oral cavity, the epithelium sits on a thick fibrous tissue sheet overlying osseous structures. In the luminal structures of the esophagus, stomach and intestine, a bilayered smooth muscle lamina is adjacent to it.

For the scientist who wants to establish an epithelial culture from the gastrointestinal tract, anatomical information where the cells are derived from is essential as this may affect the physiological function.

This chapter describes the isolation and culture techniques for oral mucosa cells.

Explant Method

- Wash fresh mucosa sample extensively in saline and rinse briefly in 70% ethanol and/or soak in 2% Penicillin–Streptomycin (100 U/ml and 100 μg/ml) in PBS
- Cut sample into 0.5 × 0.5 cm pieces
- Transfer samples into culture flasks or petri dishes containing just enough medium to wet the entire vessel surface, so as not to disturb the tissue during transportation
- Distribute tissue pieces over the culture surface, approximately 2–3 cm apart, to maximise space

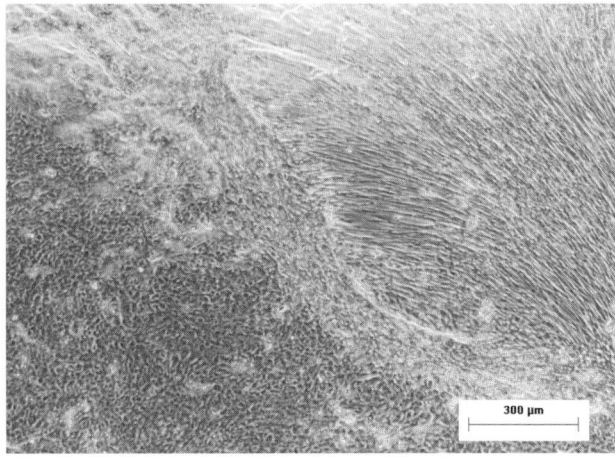

Fig. C.11a Outgrowth of fibroblasts and epithelial cells from a mucosa explant ×40.

- Leave the culture undisturbed in the incubator for 3–5 days before the first media change
- Choice of medium depends on cell population desired:
  * Epithelial cells: Defined Keratinocyte serum-free Growth Media with supplements
  * Co-culture of fibroblasts and epithelial cells: DMEM (high glucose) and Ham's F–12 in the ratio 3:1, with supplements as listed in "Requirements".

Subculturing

- At confluence, aspirate medium and incubate with 0.02% EDTA for 5 mins or longer at 37°C to remove fibroblasts. (If need be, incubate with 0.05% trypsin for a few minutes to remove confluent fibroblast sheets)
- If fibroblasts cultures are desired, neutralise aspirated trypsin–EDTA with complete DMEM, centrifuge at 200 g for 10 mins, resuspend, count and replate
- To harvest epithelial cells after fibroblasts have been removed, introduce fresh 0.1% trypsin for 5 mins at 37°C. (Leave longer if necessary)

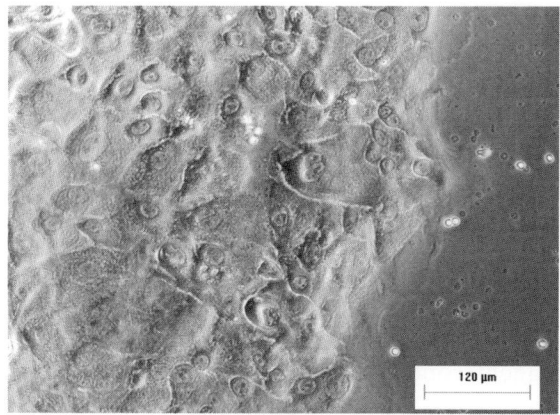

Fig. C.11b Oral mucosa epithelial cells in culture ×100.

- Check under the microscope to ensure that all cells have detached
- Tap lightly on the sides of the flasks to dislodge detaching cells
- Neutralise trypsin with complete DMEM, centrifuge at 200 g for 10 mins, resuspend, count and replate at minimum 12,000 cells per cm^2

Requirements

1. 70% ethanol
2. 2% Penicillin–Streptomycin in PBS
3. Media: DMEM (low Ca^{2+}), Ham's F–12, Defined Keratinocyte Growth Media and proprietary supplement (Gibco, NY, USA)
4. Supplements for co-culture, to be added to 3:1 DMEM and Ham's F–12:
   - 10% FBS
   - 1% Penicillin–Streptomycin
   - 10 ng/ml EGF
   - 5 μg/ml Insulin
   - 0.4 μg/ml Hydrocortisone
   - 10^{-10} M Cholera Toxin

- 20^{-12} M Triiodothyronine
- 0.18 mM Adenine
- 5 µg/ml Transferrin
5. 0.02% EDTA
6. 0.05% and 0.1% trypsin

C.3.3 Keratinocytes

Culturing epithelial cells such as epidermal keratinocytes requires more stringent control of culture medium and supplements. These cells proliferate as 2D monolayers with maximum cell-to-cell contact, giving rise to the typical "cobblestone" morphology commonly described in the literature. General techniques of isolating and culture of keratinocytes from a skin biopsy are described in this section.

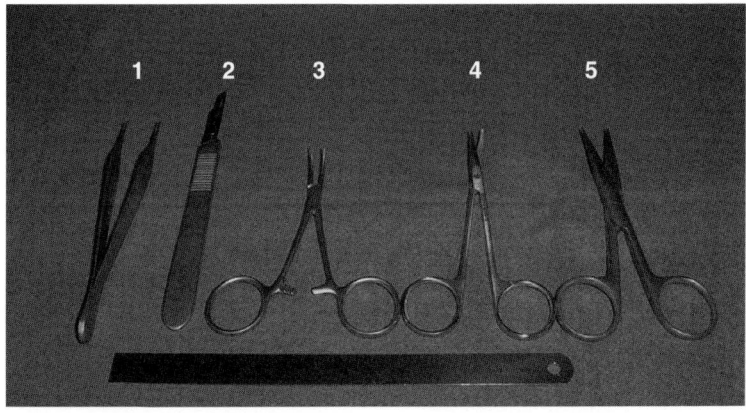

Fig. C.12a Surgical equipment for skin specimens.

Equipment Required

1. Delicate dissecting forceps
2. Scalpels with blade size 11&15
3. Dissecting and ligature clamp
4. Sharp pointed delicate curved scissors
5. Surgical straight scissors

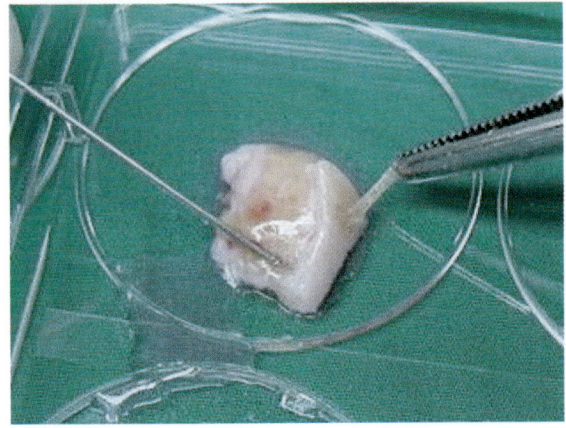

Fig. C.12b Peeling off epidermis from dermis after dispase treatment.

Separating the Epidermis from Dermis

- Wash fresh skin sample extensively in saline and rinse shortly in 70% ethanol and/or 2% Penicillin–Streptomycin (100 U/ml; 100 μg/ml) in PBS
- Cut skin sample into 1 × 0.5 cm pieces
- Immerse not more than 10 pieces of cut skin per well in a 6-well plate, into 5 mg/ml Dispase in serum-free DMEM, for 16 hrs at 4°C
- Peel off epidermis from dermis gently using sterile forceps

Isolating Single Cells from the Epidermis

- Rinse epidermis in PBS and transfer into 0.1% trypsin for 10 mins at 37°C
- Vortex well to obtain single-cell suspension and filter through 100 μm nylon sieves to remove tissue debris
- Centrifuge at 200 g for 10 mins, and plate accordingly on feeder layer or in serum-free media

Culturing on a Fibroblast Feeder Layer

- 3T3 murine fibroblasts or aulogous/allogeneic fibroblasts can be used as feeder

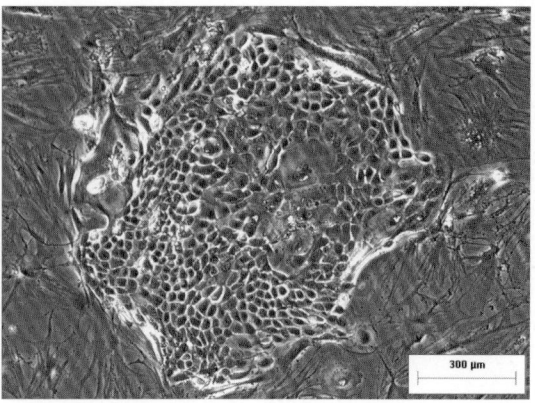

Fig. C.12c Keratinocyte colony proliferating on autologous feeder layer ×40.

- Post-mitotic treatment with Mitomycin C with following protocol
  * 3T3 fibroblasts: Plate at 8000/cm^2, treat with 4 μg/ml Mit. C in complete DMEM for 2 hrs at 37°C. Rinse 3 times with PBS and re-plate at desired density (4–20,000/cm^2) for keratinocyte seeding. Use 3:1 DMEM-Ham's F–12 with supplements as stated under "Requirements"
  * Autologous/allogeneic fibroblasts: Plate at 8000/cm^2, treat with 10 μg/ml Mit. C in complete DMEM for 2.5 hrs at 37°C. Rinse 3 times with PBS and re-plate at desired density (4–20,000/cm^2) for keratinocyte seeding. Use 3:1 DMEM–Ham's F–12 with supplements as stated under "Requirements"
- Plate keratinocytes in desired density (1–20,000/cm^2) on feeder layer
- First medium change after at least 3 days
- Maintain post-mitotic fibroblasts in DMEM with 10% FBS and 1% Penicillin–Streptomycin until use
- Discard cells after 1 month if not used

Culturing in Serum-free Medium

- Plate isolated keratinocytes at density of at least 15,000/cm^2 in serum-free Defined Keratinocyte Medium with supplement

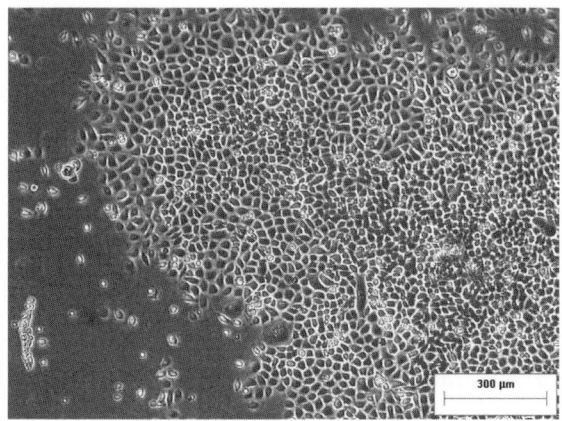

Fig. C.12d Keratinocyte colony proliferating in serum-free medium ×40.

- Leave culture flasks undisturbed for at least 3 days before first medium change
- If cells do not attach, wait for a few more days before medium change

Subculturing for both Feeder-layer and Serum-free

- Aspirate medium, incubate with 0.02% EDTA for 1 min at 37°C
- For feeder layer system, incubate for 5 mins or longer to remove remaining feeder cells. (If need be, incubate with 0.05% trypsin for a few minutes to remove feeder cells)
- Aspirate EDTA and add 0.1% trypsin for 5 mins at 37°C. (Leave longer if necessary)
- Neutralise trypsin with complete DMEM, centrifuge at 200 g for 10 mins, resuspend, count and replate.

Requirements

1. 70% ethanol
2. 2% Penicillin–Streptomycin in PBS
3. 5 mg/ml Dispase in serum-free DMEM (sterile filtered)
4. Media: DMEM (high glucose, low Ca^{2+}), Ham's F–12, Defined Keratinocyte Growth Media and proprietary supplement (Gibco, NY, USA)

5. Supplements for feeder layer culture system, to be added to 3:1 DMEM and Ham's F–12:
   - 10% FBS
   - 1% (50 U/ml; 50 µg/ml) Penicillin–Streptomycin
   - 10 ng/ml EGF
   - 5 µg/ml Insulin
   - 0.4 µg/ml Hydrocortisone
   - 10^{-10} M Cholera Toxin
   - 20^{-12} M Triiodothyronine
   - 0.18 mM Adenine
   - 5 µg/ml Transferrin
6. 0.02% EDTA
7. 0.05% and 0.1% trypsin
8. 4 mg/ml or 10 mg/ml Mitomycin C (depending on feeder layer system)
9. 3T3 mouse or primary autologous/allogeneic fibroblasts

C.3.4 Hepatocytes

The liver is the second largest organ of the human body next to skin, and performs a variety of important physiological functions. The organ consists of a left, right and caudal lobe, and is situated in the upper right quadrant of the abdomen. The liver is responsible for the synthesis and secretion of proteins like albumin, fibrinogen and various other plasma factors in the coagulation cascade. In addition to that, the liver plays a central role in glucose and fat metabolism. The histomorphological structure is complex with cells organized in lobuli, centered on the Glisson–Tris consisting of the bile duct and vascular supply. Next to hepatocytes, endothelial cells, Itocells (fat-metabolism) and Kupffer-cells (phagocytosis) are present. This section focuses on the culture of hepatocytes.

Note

As primary human parenchymal tissue samples are normally quite difficult to get for biomedical research purposes, the following protocols for Hepatocytes (C.3.4) and Kidney cells (C.3.5) were derived from a tissue specimen harvested from an animal. However, the principle technologies are identical.

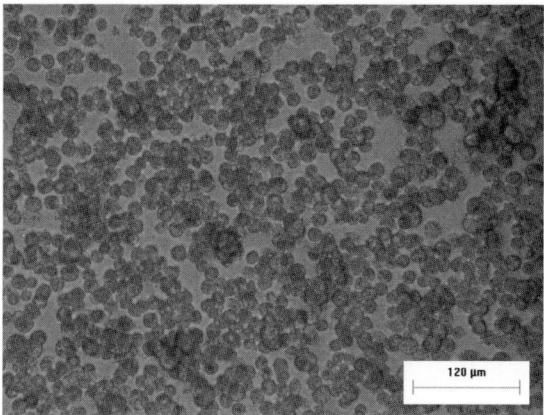

Fig. C.13 Hepatocytes in culture ×100.

Enzymatic Digestion

- Harvest hepatocytes by a two-step *in situ* collagenase perfusion method
- After laparotomy, place and fix a portal cannula in position along the portal vein
- Promptly make a cut in the lower vena cava
- In the initial first 2–3 mins, perform preperfusion with Ca^{2+}-free perfusion buffer while the liver remains *in situ*
- Start the perfusate flow at a rate of 50 ml per minute
- While preperfusion was carried out, transfer the liver to a petri dish and place in a position similar to its *in situ* site
- After 10 mins of preperfusion with Ca^{2+}-free medium, perform perfusion with recirculating 0.05% collagenase buffer for another 10 mins. Terminated when the vena cava ruptures
- Perform the entire perfusion procedure under oxygenation to improve the cell viability
- Liberate cells from the connective vascular tissue and resuspend in fresh growth medium
- Incubate cell suspension at 37°C, 5% CO_2 for 30 mins
- Filter cell suspension through 60 μm nylon membranes to further remove the connective tissue debris
- Centrifuge filtrate at 50 g for 1 min to obtain a pellet
- Resuspend and wash cell suspension twice with Ham's F–12 culture medium, supplement with 0.2% bovine albumin and 10 μg/ml bovine insulin, and seed into culture vessels

Requirements

1. Ca^{2+}-free perfusion buffer
2. 0.05% Collagenase
3. Media: Ham's F–12 supplemented with 0.2% bovine albumin and 10 μg/ml bovine insulin

C.3.5 Kidney Cells

This section focuses on cell culture of cortical tubular and mesangial cells of the kidney. The kidney is a largely heterogeneous organ situated in the retroperitoneum and consists of a parenchymatous tissue with distinct tissue architecture of an outer fibrous sheet encapsulating the organ with its cortex and an inner core region. The cortex hosts the glomeruli and proximal tubuli; and the core region, the medullary interstitium with the distal tubuli, the urether and the large vascular supply. Kidney tubule epithelial cells can be successfully cultured under serum-free conditions.

A tissue biopsy is usually obtained during nephrectomy due to renal cell carcinoma. Several techniques have been used to isolate and culture human mesangial, glomerula and tubular cells using explant or enzymatic methods or cell isolation via continuous density gradient centrifugation with Percoll.

Renal Cortical Cell Culture

This method is relatively easy to establish, but has limitations in terms of cell purity as a co-culture of tubular, mesangial and vascular (endothelial cells) might be present.

- Place the tissue specimen briefly into a beaker with 70% ethanol, followed by rinsing with physiological saline solution or PBS
- Transfer the sample to a large petri dish and remove the fibrous tissue capsule (always add drops of fresh saline solution from a syringe to prevent dehydration of the tissue)
- Mince the tissue into small pieces using a size 11 scalpel (try to avoid including vascular structures in the tissue prep)
- Add PBS and 0.25% trypsin–EDTA to the tissue prep and transfer it into a 50 ml centrifuge tube, and onto a gently shaking platform for 20–30 mins

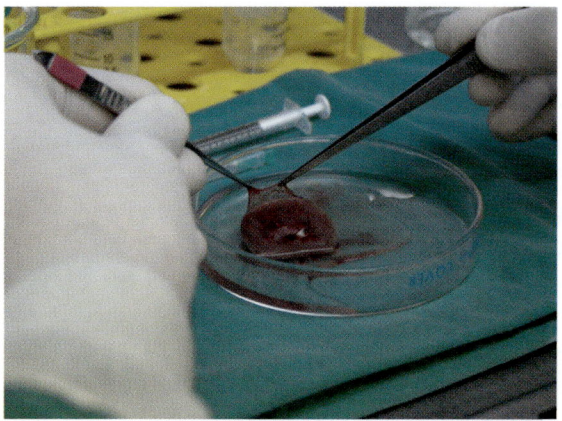

Fig. C.14 Dissecting the capsule from the parenchyma of a fresh kidney sample.

- Filter the cells through a sterile 50 μm nylon sieve
- Centrifuge the cell suspension at 200 g for 10 mins at 4°C, and discard supernatant
- Resuspend the pellet in complete media (1:1 DMEM and Hams F–12 medium supplemented with 10% FBS, 1% Penicillin, 1% HEPES and 5×10^{-8} M hydrocortisone)
- First media change after 3 days

Requirements:

1. PBS
2. 0.25% trypsin-EDTA
3. 70% ethanol
4. Media: DMEM Ham's F–12, supplemented with
   - 10% FBS
   - 1% Penicillin
   - 1% HEPES
   - 5×10^{-8} M hydrocortisone
5. 50 μm Nylon sieves

Proximal Tubuli Cell Culture

The method described here is a combination of microdissection and enzymatic digestion, followed by gradient centrifugation, which results in a more homogeneous culture as compare to the previously described method.

- Place the tissue specimen briefly into a beaker with 70% ethanol, followed by rinsing with physiological saline solution or PBS
- Transfer the sample to a large petri dish and remove the fibrous tissue capsule — always add drops of fresh saline solution from a syringe to prevent dehydration of the tissue
- Mince the tissue into small pieces using a size 11 scalpel (try to avoid including vascular structures in the tissue prep)
- Depending on the amount of tissue add 0.25% trypsin (alternatively 0.2% collagenase type IV) and transfer the solution into a culture vessel and incubate it at 37°C for 20–30 mins
- Filter the solution through a 300 μm nylon sieves, followed by rinsing the solution once with fresh complete medium. Centrifuge the cell-tissue solution at 200 g and filter it again through a 100 μm nylon sieve
- Centrifuge the cells again and resuspend the pellet in 10 ml of complete medium
- Aliquot two 5 ml tubes and overlay the cell suspension onto a 30 ml Percoll solution
- Centrifuge at 420 g in a fixed angle rotor for 30 mins
- Isolate the lowest band sitting on a percoll gradient close to the bottom of the tube
- Resuspend the cells in complete medium and transfer into culture flasks

Requirements

1. 0.2% Collagenase type IV dissolved in media, activity 0.5 U/mg
2. 0.25% trypsin
3. Fixed-angle rotor centrifuge
4. Nylon sieves (100 μm and 300 μm)

Glomerula Mesangial Cell Culture

The glomerula of the renal cortex consists of four distinct cell types: epithelial, contractile mesangial, bone marrow derived mesangial and endothelial cells. However this histomorphological compartmentalization is lost in cell culture.

The easiest and most straight forward method of culturing those cells is by explant culture of isolated glomerula. Cobblestone-like epithelial cells will dominate the culture next to spindle-shaped mesangial cells spreading radially from the glomeruli.

If the experiment requires a homogeneous population of one of the glomerula cells, antibody-labelled surface-markers can be used.

- After manually mincing the cortex of the tissue sample (see protocol above), resuspend the tissue fragments and pass them through a 300 μm, 200 μm, and finally, a 100 μm nylon sieve
- The glomeruli will be retained on top of the 100 μm sieve and are resuspended and cultured in RPMI 1640 medium supplemented with 10% FBS, 1% ITS 1 (**I**nsulin, **T**ransferrin, **S**odium Selenite) and 1% Penicillin

Requirements

1. Media: RPMI 1640, supplemented with
   - 10% FBS
   - 1% ITS 1 (available from Sigma)
   - 1% Penicillin

Note

The medium used should not be supplemented with streptomycin, which is a nephrotoxin!

C.4 Bone Marrow Derived Cells

Several bone marrow derived cells, in particular haematopoietic cells but even some precursor mesenchymal stromal cells, can be grown in suspension or attached on a culture plate. The cells usually form clusters in culture Colony Forming Units (CFU), which are density dependent. Haematopoietic cells can be cultured under serum or serum-free conditions. Haematopoietic cells are usually isolated via gradient centrifugation method or via Fluorescence Activated Cell Sorting (FACS).

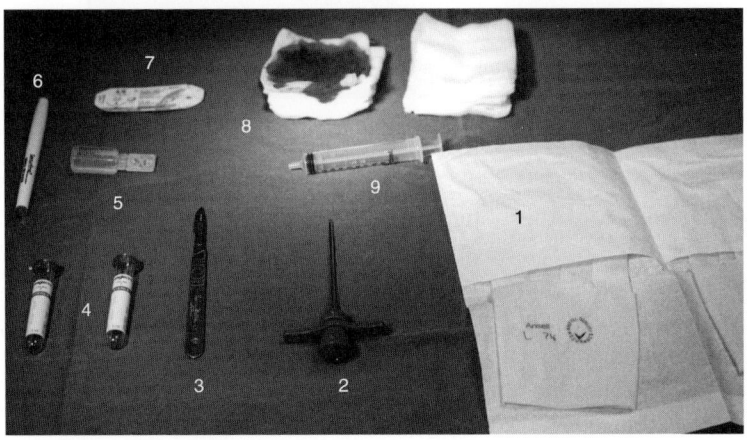

Fig. C.15 Surgical equipment for bone marrow aspiration.

Equipment Required

1. Sterile surgical gloves
2. Bone marrow aspiration needle
3. Scalpel for stab incision
4. Coated heparinised tubes
5. Heparin to coat syringe
6. Marker
7. Suture
8. Gauze
9. Syringe

C.4.1 Bone Marrow Stromal Cells

- Bone marrow samples are aspirated and collected into vacuum polypropylene tubes containing Heparin and mixed well to prevent clotting

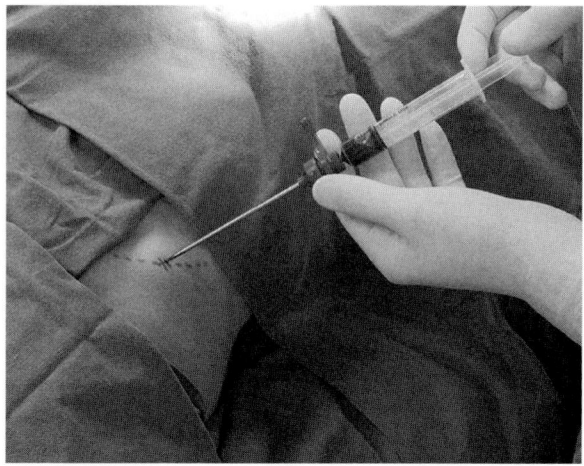

Fig. C.16a Aspirating bone marrow from the anterior iliac crest.

- Measure out the bone marrow sample using sterile pipettes and place it in a 50 ml centrifuge tube
- Add an equal amount of PBS to the same tube and centrifuge at 900 g for 10 mins and discard supernatant
- Homogenise the pellet with 20 ml PBS
- Use 50 ml of the prepared PERCOLL (43.85 ml of PBS: 56.15 ml of Percoll — 1.13 density) and place 25 ml per 50 ml tube
- CAREFULLY place 10 ml of the homogenised solution into each of the two tubes, on the 25 ml of PERCOLL as an unmixed overlay. (Do this very slowly using a pipette)
- Ensuring that the fluid interface DO NOT MIX, centrifuge at 900 g for 30 mins at 22°C.
- Obtain the light pink ring of cells that is sandwiched between the upper yellowish layer of PERCOLL and the red blood cells layer. (Use a pipette and do this slowly)
- Place the fluid layer in a new 50 ml centrifuge tube
- Centrifuge at 200 g for 10 mins at 22°C
- Obtain the pellet and discard supernatant. (Be careful since the pellet is usually not obvious; take the reddish fluid at the base of the tube)
- Place the pellet in a 10 ml centrifuge tube and re-homogenise with 5 ml of PBS
- Centrifuge at 200 g for 10 mins at 22°C and discard supernatant
- Homogenise pellet with 1 ml of media and count the cells

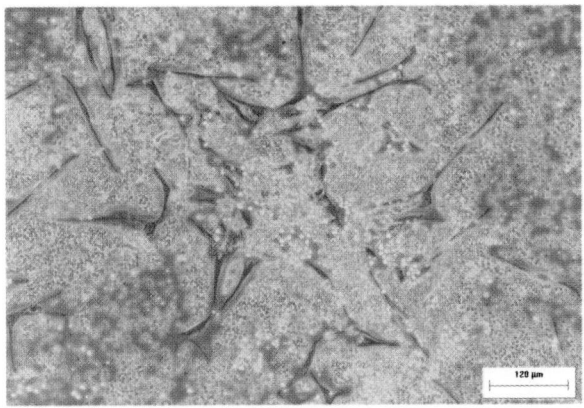

Fig. C.16b Bone marrow stromal cells in culture ×100.
Non-adherent small-sized haematopoietic cells are washed
away with the first media change.

- Transfer the cells into a 25 cm² culture flask and culture
- First media change after 3 days

C.4.2 Peripheral Blood Mononuclear Cells

Isolating Peripheral Blood Mononuclear Cells from blood with
density gradient centrifugation.
- Start with freshly drawn human blood or buffy coat, not older
  than 8 hrs, treated with an anticoagulant
- Dilute cells with 2–4 volumes of PBS (the more dilute the blood
  sample, the better the purity of the mononuclear cells)
- Carefully layer 35 ml of diluted cell suspension over 15 ml
  Ficoll Paque (1.077 density) in a 50 ml conical tube and
  centrifuge at 400 g for 30 mins at 20°C in a swinging bucket
  rotor (without brake)
- Aspirate the upper layer, leaving the mononuclear cell layer
  undisturbed at the interphase
- Carefully transfer the interphase cells (lymphocytes,
  monocytes, thrombocytes) to a new 50 ml conical tube
- Fill the conical tube with PBS, mix and centrifuge at 300 g for
  10 mins at 20°C. Carefully remove the supernatant completely
- For removal of platelets, resuspend the cell pellet in 50 ml of
  buffer and centrifuge at 200 g for 10–15 mins at 20°C. Carefully

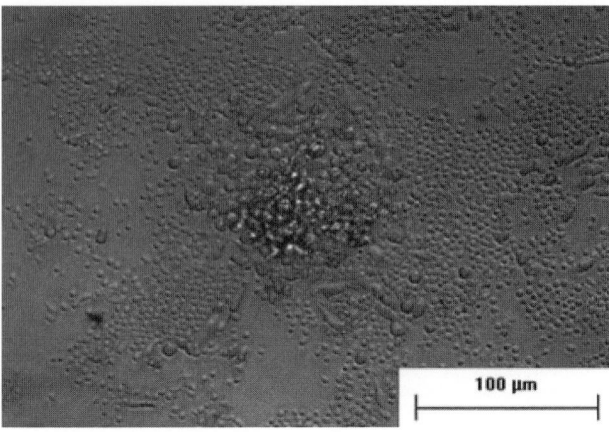

Fig. C.17 Peripheral blood mononuclear (PBMC) cell culture ×100.

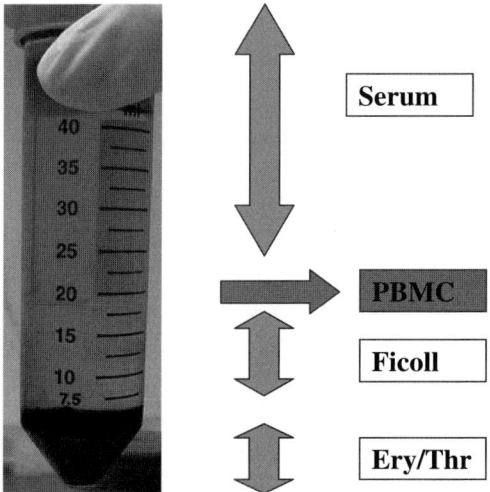

Fig. C.18 Isolation of pheripheral blood mononuclear cells using density gradient centrifugation over a Ficoll (Amersham Pharmacia Biotech™) layer. Mononuclear cells accumulate in the white opaque interphase between the serum fraction and the Ficoll-pague.

Note: the fraction contains ~ 90% lymphocytes and ~ 10%. monocytes. To further separate the monocytes from the lymphocytes, the cells are transferred into a culture dish. Monocytes adhere to the dish within 4–6 hours whereas lymphocytes can be washed away.

remove the supernatant completely. Repeat this last washing step. Most of the platelets will remain in the supernatant upon centrifugation at 200 g

• Resuspend the cell pellet in buffer. Count the cells, resuspend them in complete media and plate cells into culture dishes

Requirements

1. Buffer (degassed): PBS pH 7.2, supplemented with 0.5% BSA & 2 mM EDTA. Keep buffer cold (4–8°C)
2. Ficoll–Paque SG 1.077 and PBS supplemented with 2 mM EDTA
3. PBS
4. 50 ml conical tubes

Note

Peripheral blood mononuclear cells and bone marrow derived stromal cells are plating-density-sensitive cells. It is therefore advised (depending on the number of initially isolated cells) to start the culture in T25 flasks or 60 mm cell culture dishes.

D

**Reagents
and Techniques
for Cell Culture**

Media is the most important factor in culturing cells *in vitro*. The function of the media is to provide nutrients and to maintain physiological conditions in respect to pH, CO_2, O_2 and osmotic balance.

D.1 Selection of Cell Culture Media

D.1.1 Classical Media

Defined cell culture media generally consists of 4 basic chemical groups: amino acids, carbohydrates, inorganic salts and vitamins. This balanced mixture provides the sources for protein synthesis, energy metabolism, cell membrane functions and catalytic processes.

Media can be either in liquid or powder form. Very often phenol red is added as an indicator for the pH. What the various changes in colour indicate is discussed in the troubleshooting section. CO_2 is generally required for maintaining the pH. However, special HEPES-buffered media can be used under CO_2-free conditions. Choosing the 'right' media from a large variety of formulations is not always easy; many primary human cells grow in a number of media preparations. It again depends on the personal preference. Basic recommendations are given here. Publications in the field serve as a useful reference when choosing the right media to use. Frequently used media preparations are mentioned in this section:

Minimum Essential Media (MEM)

- Supports the growth of a broad spectrum of cells (including haematopoietic and bone marrow cells)
- Most widely used media
- Originally for primary cultures
- Basis for modification (DMEM, GMEM, …)

Dulbecco's Modified Eagle Media (DMEM)

- Well suited for a broad spectrum of mammalian cells
- Can be used as a basic component for speciality media
- Modification of MEM
- Increased amino acids and vitamins

- Non-essential amino acids
- Ferric nitrate
- More glucose

Nutrient Mixture F–12 (Hams' Modification)

- Well suited for chondrocytes, mucosa cells
- Higher level of amino acids, vitamins and trace elements
- Developed for cell lines but suitable for primary cells

Roswell Park Memorial Institute (RPMI 1640) Media

- Modification of the Mc Coy's 5a media for hematopoietic cells in culture
- Supports a broad spectrum of cells
- Often used for hybridoma cultures

Mc Coy's 5a Media

- Increased levels of glucose, vitamins and peptone
- Suitable for cells from liver, intestine, bone marrow and gingiva biopsies

Media 199

- Suitable for osteoblast cultures
- Relatively high Ca content
- First chemically defined media widely used in virology & primary explants

Note

- Storage: 4°C in the dark
- Add: Serum, anti-microbial agents and supplements such as L–glutamine prior to use
- Aliquot samples in suitable volumes

D.1.2 Serum-free and Speciality Media

Speciality media are specially formulated media to grow specific cell types or perform a specific application. The biggest group of speciality media are the serum-free ones. Although without serum, these media may contain specific, discreet proteins or

Table 1 Formulations of common cell culture media (liquid).

	MEM (Eagles)	DMEM	Hams F-12	RPMI 1640	M199	McCoy
Components (mg/L)						
Salts						
$CaCl_2$	222.00	200.00	33.20		200.00	100.00
$Ca(NO_3)$ $4H_2O$				100.00		
$Fe(NO_3)_3$ $9H_2O$		0.10			0.72	
$FeSO_4$			0.83			
KCL	400.00	400.00	223.60	400.00	400.00	400.00
KH_2PO_4					60.00	
$MgSO_4$	98.00	97.67		48.84	98.00	98.00
$MgCl_2$			57.22			
NaCl	6800.00		7599.00	6000.00	6800.00	5100.00
$NaHCO_3$	2200.00	3700.00	1176.00	2000.00	2200.00	2200.00
NaH_2PO_4 H_2O	140.00	125.00	142.00	800.00	140.00	580.00
ZnSO4			0.86			
$CuSO_4$ $5H_2O$			0.0025			
Others						
Adenine Sulfate					10.00	
Adenosine 5-triphosphate					1.00	
Adenosine 5 Phosphate					0.20	
Cholesterole					0.20	
2 Deoxy Ribose					0.50	
Deoxyribose						
D-Glucose	1000.00	4500.00	1802.00	2000.00	1000.00	3000.00
Guanine					0.30	
Phenol Red	10.00	15.00	1.20	5.00	20.00	10.00
HEPES						5958.00
Hypoxanthine			4.77		0.40	
Ribose					0.50	
Sodium Acetate					50.00	
Sodium Pyruvate		110.00	110.00			

Table 1 (Cont.)

	MEM (Eagles)	DMEM	Hams F-12	RPMI 1640	M199	McCoy
Bacto-Peptone						600.00
Succinic Acid						
Linoleic Acid			0.08			
Lipoic Acid			0.21			
Putrescine 2HCL			0.161			
Polysorbate					20.00	
Thymidine			0.70		0.30	
Uracil					0.30	
Xantine Na					0.34	
Glutathione				1.00	0.05	0.50
Aminoacids						
L-Alanine	9.00		8.90		25.00	13.90
L-Argine HCL	126.00	84.00	211.00	200.00 (HCLfree)	70.00	42.10
L-Aspartic Acid	13.00		13.00	20.00	30.00	20.00
L-Asparagine	13.00		15.00	50.00 (freebase)		45.00
L-Cystine 2 HCL	31.00	63.00	35.00	65.00	26.00	31.50
L-Glutamic Acid			14.70	20.00	75.00	22.10
L-Glutamine (acid)	15.00		146.00	300.00	100.00	219.20
Glycine	8.00	30.00	7.50	10.00	50.00	7.50
L-Histidine HCL	42.00	42.00	21.00	15.00	22.00	21.00
L-Hydroxyproline				20.00	10.00	19.70
L-Isoleucine	52.00	105.00	4.00	50.00	40.00	39.40
L-Leucine	52.00	105.00	13.00	50.00	60.00	39.40
L-Lysine	72.50	146.00 (HCL)	36.50	40.00	70.00	36.50
L-Methionine	15.00	30.00	4.50	15.00	15.00	15.00
L-Phenylalanine	32.00	66.00	5.00	15.00	25.00	16.50
L-Proline	12.00		34.50	20.00	40.00	17.30
L-Serine	11.00	42.00	10.50	30.00	25.00	26.30
L-Threonine	48.00	95.00	12.00	20.00	30.00	17.90
L-Thryptophan	10.00	16.00	2.00	5.00	10.00	3.10

Table 1 (Cont.)

	MEM (Eagles)	DMEM	Hams F-12	RPMI 1640	M199	McCoy
L-Thyrosine	52.00 (2Na2H2O)	104.00	7.80	29.00	58.00	26.20
L-Valine	46.00	94.00	11.70	20.00	25.00	17.60
Vitamins						
Ascorbic Acid					0.05	0.50
α-Tocopherol Phosphate					0.01	
Biotin			0.007	0.20	0.01	0.20
Calciferol					0.10	
D-Ca Pantothenate	1.00	4.00	0.50	0.25	0.01	0.20
Choline Chloride	1.00	4.00	14.00	3.00	0.50	0.20
Folic Acid	1.00	4.00	1.30	1.00	0.01	10.00
I-Isonitol	2.00	7.20	18.00	35.00	0.05	36.00
Menadione					0.01	
Niacin					0.025	0.50
Niacinamide	1.00	4.00	0.04	1.00	0.025	0.50
Para-aminobenzoic Acid				1.00	0.05	1.00
Pyridoxal HCL	1.00		0.06	1.00	0.025	0.50
Pyridoxine HCL		4.00			0.025	0.50
Riboflavin	0.10	0.40	0.04	0.20	0.01	0.20
Thiamin HCL	1.00	4.00	0.03	1.00	0.01	0.20
Vitamin A (acetate)					0.14	
Vitamin B12			1.40	0.005		2.00

bulk protein fractions. This allows for a more controlled set of conditions, which ensures more consistent performance and therefore more precise evaluation of cellular functions.

To achieve optimal growth of the cultures it is until now often necessary to add serum from animal or human to the cell culture media. But the addition of sera is connected with some problems.

Disadvantages of Using Sera

1. For most cells serum is not the physiological fluid
2. Contents of ingredients vary from lot to lot
3. Sera could contain inhibitors or toxins
4. Some ingredients could cause unwanted reactions
5. Serum is a source of contamination
6. In primary cultures, the growth of fibroblasts is promoted by sera

Advantages of Serum-free Media

1. Avoids qualitative and quantitative fluctuations of nutrients
2. Growth conditions are defined and controlled
3. Small chances of contamination
4. Serum-free media allows the possibility of studying the effects of one component present in serum, while eliminating the rest.
5. Isolation and characterisation of expressed proteins is easier

General Considerations

- Dis-/advantages depends on the type of cell and their intended purpose
- Defined media could also contain inhibitors
- Some serum-free media are expensive

Serum-free media can be separated into **P**rotein-**F**ree **M**edia (PF Media) and **C**hemically **D**efined **M**edia (CD Media). PF media contains no intact proteins but hydrolysated animal or plant derived proteins (peptone). CD media contains a few totally characterised proteins, which should mimic the serum-effects. The proteins are animal-derived or microbially-derived recombinant proteins. A few of speciality media are briefly described here.

Human Endothelial Serum Free Media

- Supports isolation and expansion of human arterial and venous endothelial cells
- To be supplemented with bFGF, EGF and human plasma fibronectin

Keratinocyte Serum Free Media

- Optimised for cultivation of primary and passaged human keratinocytes
- Retards fibroblast growth
- Low calcium level encourages proliferation and delays differentiation

Opti-MEM Media

- Modification of Eagle's MEM, allowing at least a 50% reduction in serum supplement
- Ideal for culturing cells during cationic lipid transfections

D.2 Cell Culture Reagents/Media supplements

D.2.1 Sera

Serum is the fraction of blood remaining after coagulation and centrifugation of the cellular part. It is an extremely complex mixture of plasma proteins, growth factors, hormones etc.

The various components may vary based on the origin and nutritive conditions; therefore it is important to note that there can be substantial batch to batch variations. Usually **F**etal **B**ovine **S**erum (FBS) is used for cell culture. Besides the FBS, other sera are **N**ew **B**orn **C**alf **S**erum (NBCS), bovine/horse serum (from slaughtered animals) and donor bovine/horse serum (from living animals; better quality). There are also **S**ynthetic **S**erum **S**ubstitutes (SES) as well as various other animal sources. Depending on the processing, there are different kinds of sera:

1. Heat pre-inactivated sera
2. Low lgG sera
3. Dialysed sera
4. Gamma-irradiated sera

Finally, as serum is probably one of the most costly parts in cell culture, deciding which serum to choose depends largely on your lab director's financial status!

- Storage: −20°C
- Heat deactivate: +60°C, 30 mins prior to use
- Aliquot 50 ml portions or other volumes as required
- No repeated thawing and freezing
- Slow thawing

Protocol for Isolation of Serum from a Blood Sample

- Withdraw whole blood using large syringes; do not use any anticoagulation factors (EDTA/Heparin) as this leads to clotting of the erythrocytes with the clotting factors
- Centrifuge the blood at 5000 g at 4°C for 20 mins
- Take off the serous supernatant and heat deactivate at 60°C for 30 mins
- Filter through 0.2 μm nylon filter, aliquot and freeze at −20°C

Important

> Flocculence in serum — most commonly denatured lipo-
> proteins or fibrin clots. Usually this does not affect the cell
> culture.

D.2.2 Antibiotics (AB) and Antimycotics (AM)

Cultures can be contaminated with bacteria and/or fungus when
not handled properly, or when the working environment is not
sterile. The decision to use antibiotics and/or antimycotics to
prevent contamination should be based on the individual
researcher's need and experience.

There is no general rule for using antibiotics (AB) and
antimycotics (AM) and establishing a primary cell line often
depends on the conditions and the anatomical location of your

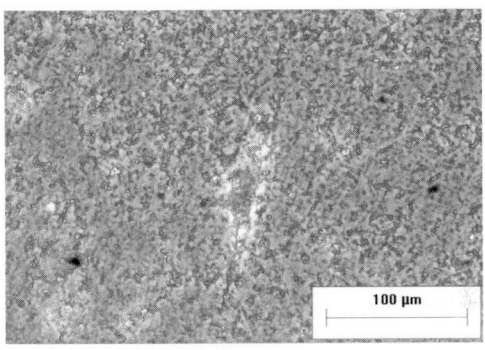

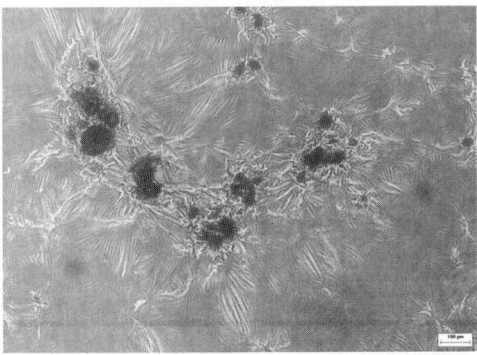

Fig. D.1 *Top*: Bacteria contaminated culture (×400). *Bottom*: Fungus
contaminated culture (×100). Note the difference in size!

tissue samples. Many primary cultures contain some bacteria in low concentrations and that is generally nothing to worry about. Table 2 is a general guide and the concentrations are given for tissue culture media containing serum. Serum-free media generally requires lower AB/AM concentrations. However, the optimal concentration should be determined empirically.

Finally, it is important to state that even though AB/AM is used, proper handling and sterile culture techniques are still necessary. Table 2 lists the most commonly used antibiotics and antimycotics for cell culture.

Note

1. Storage: room temperature, 4°C or −20°C, depending on the AB/AM
2. Aliquot 5/10 ml portions, or in other volumes as required

When a culture becomes contaminated, there are a number of steps to control the problem.

1. Determine the contaminant: fungus, yeast, bacteria, mycoplasma, virus
2. Isolate the culture from another culture — place on another shelf of the incubator
3. Wash the culture at least 3 times with warm PBS, then add fresh media
4. Change media containing AB/AM, depending on the contaminating micro organism, on a daily basis
5. Change to another AB/AM as those used might not cover the spectrum, or a resistant strain might have been selected
6. If you are unable to control the contamination after 4 days, discard the culture

D.2.3 Biological Buffers

Biological buffers or balanced salt solutions (BSS) compose of inorganic salts and, in some cases, sodium bicarbonate and glucose. These buffers serve as basis for most culture media, preventing undesirable fluctuations in pH, which can be detrimental to the cells. Most cells require pH conditions in the

Table 2 Commonly used AB/AM for cell culture.

Antibiotic	Antimicrobial spectrum	Recommended concentration (μg/ml media)	Stability in media at 37°C (days)
Amoxicillin	Gram-positive and gram-negative bacteria	100	3
Ampicillin	Gram-positive and gram-negative bacteria	100	3
Erythromycin	Gram-positive bacteria and Mycoplasma	100	3
Penicillin V	Gram-positive bacteria	100	3
Tetracycline	Gram-positive and gram-negative bacteria and mycoplasma	10	4
Amphotericin B	Fungi and yeasts	0.25–2.5	3
Gentamicin Sulphate	Gram-positive and gram-negative bacteria and mycoplasma	5–50	5
Kanamycin Sulphate	Gram-positive and gram-negative bacteria and mycoplasma	100	5
Neomycin Sulphate	Gram-positive and gram-negative bacteria	50	5
Nystatin	Fungi and yeasts	100 (units/ml)	3
Penicillin G	Gram positive bacteria	50–100 (units/ml)	3
Polymixin B Sulphate	Gram negative bacteria	100 (units/ml)	5
Streptomycin Sulphate	Gram-positive and gram-negative bacteria	50–100	3

Table 3 Commonly used buffers in cell culture.

Buffer/ Salt solution	Characteristics and functions
HEPES	Most effective buffer in the pH range 7.2–7.8.
PBS	Common diluent and rinsing medium. PBS Solution A is PBS without Ca^{2+} and Mg^{2+}. Useful for reconstituting enzymes for tissue dissociation or lifting cells from substrates.
HBSS	Suitable for use in sealed culture flasks incubated in air. Contains glucose.
Tris	*In vitro* buffer for assay reagents, has low metal content. Protection against preservation of enzyme activity.
Earle's BSS	Higher bicarbonate concentration compatible for cell growth in 5% CO_2. Contains glucose.

range of 7.2–7.4. There are however variations, with fibroblasts prefering a higher pH (7.4–7.6) and transformed cell lines prefering acidic conditions (7.2–7.4). Buffers are also used as diluents of amino acids, minerals and vitamins as supplements for culture media, and also for reconstituting enzymes for tissue dissociation. Various kinds of buffers are available and can be confusing to a cell culture novice. The best reference on which buffers to use for different purposes is still the relevant publications in the particular field. A few common buffers and their characteristics are briefly described in Table 3.

Sodium bicarbonate ($NaCO_3$) is the major buffer component in media as well as an essential nutrition. In water, sodium bicarbonate dissociates into Na^+ and CO_2. For maintaining the right pH-value, the buffer system has to be supplied with CO_2.

The effect of the buffer is connected to the CO_2 tension

Dissociation of $NaHCO_3$:

$$NaHCO_3 + H_2O \rightleftharpoons Na^+ + H_2O + HCO_3^- \rightleftharpoons Na^+ + OH^-$$
$$+ H_2CO_3 \rightleftharpoons Na^+ + H_2O + CO_{2(g)} + OH^-$$

Increase of H^+ : $H^+ + HCO_3^- \rightleftharpoons H_2CO_3 \rightleftharpoons CO_2 + H_2O$

Increase of OH^- : $OH^- + H_2CO_3 \rightleftharpoons HCO_3^- + H_2O$

Besides serum, antibiotics/antimycotics buffer and base salt solution, complete media contains a whole spectrum of supplements including essential and non-essential amino acids, lipids, vitamins, salt, glucose and growth factors. Various media formulations have been developed based on different combinations of these supplements to tailor to growth characteristics of different cell types or to induce specific cellular responses, such as differentiation. The researcher can achieve his desired purpose by adding further specialised supplements. Because of the plethora of supplements available, a description of individual ingredients is not possible here. Those interested in finding out more should turn to commercial catalogues and reference texts.

D.2.4 Supplements

Water

The quality of the media depend to a large extend on the purity of the water. There are various procedures for preparing pure water from tap water.

- Distillation
- Deionisation
- Reverse osmosis
- Adsorption
- Filtration
- UV Light

L–Glutamine

L–Glutamine is unstable at temperatures above $-10°C$ due to thermal or enzymatic break up. A lack of this essential amino acid can be avoided by changing the medium or by using L–Glutamine containing dipeptides (e.g. Glutamax®). These dipeptides are stable at physiological conditions. The cells release the L–Glutamine on demand by enzymatic break-up. The use of Glutamine dipeptides enhances the life time of the media and avoids inconsistency due to uncontrolled glutamine degradation.

Lipids

Lipids are an important energy source (triglyceride); they are essential for building up the cell membrane (phospholipids, cholesterol) and function as growth factor or signal molecules (steroids). Lipids are contained in the basal media or can be supplemented as concentrates (e.g. Gibco™ cholesterol, lipids concentrates).

D.3 Cell Dissociation Products

To perform a passage of a confluent monolayer cell culture or to isolate cells from a tissue sample, dissociation and detachment of the cells are required. This is achieved by breaking up the extra cellular matrix (ECM) of the cell. The final goal is to maximise the yield of viable, single cells. The success of the dissociation and detachment of the cells is determined by nine parameters:

- Type of the tissue
- Species of the origin
- Age of the species
- Dissociation medium
- Used enzymes
- Impurities in enzyme solution or medium
- Final concentration of the enzyme
- Incubation temperature
- Incubation time

The structure of the ECM depends on the cell type and is very complex and dynamic. So there is no general procedure that works for all cell cultures but there is a general strategy of introducing or optimising procedures for cell dissociation.

D.3.1 General Considerations

To achieve high yields harsh conditions (high concentration of enzyme, long incubation times etc.) are required, but the harsh conditions lower the viability of the released cells. So there has to be a negotiation between cell yield and viability of the cells.

Low Yield and Low Viability

Conditions are too harsh. Change to less digestive type of enzyme and/or decrease in working concentration to improve the results.

Low Yield and High Viability

Dissociation is too little. Choose higher enzyme concentration or incubation time. If yield remains poor, use a higher digestive enzyme.

High Yield and Low Viability

Good release of the cells but cellular damage occurs. Use lower enzyme concentrations and/or incubation times. Otherwise, use a less proteolytic enzyme.

High Yield and High Viability

Record the effects of changing dissociation parameters to learn their limitations for future optimisations.

D.3.2 Basic Strategy

Based upon the cited and referenced working concentration of the enzyme(s), prepare a preliminary dissociation batch. If more than one enzyme is used, optimise the concentration of the enzyme with the highest relative concentration (primary enzyme) before adding the secondary enzyme.

Vary the recommended concentration by approximately 50%. Prepare a dilution series with evenly distributed concentrations that cover the entire range. The quantitation of the yield and the viability is performed by counting the cells in a counting chamber after trypan blue staining.

When the optimal concentration is found, further optimisation is reached by varying incubation temperature and time.

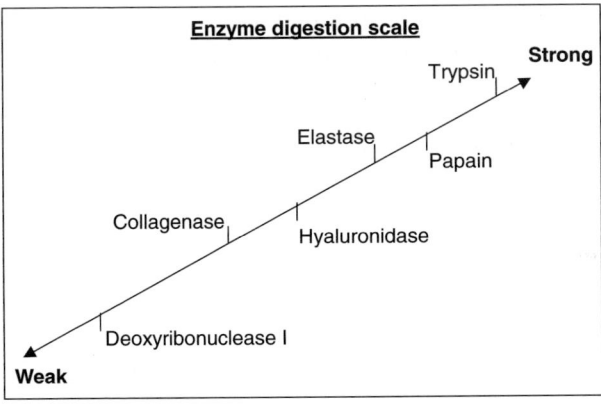

Fig. D.2 Enzyme digestion scale.

D.3.3 Trypsin

Trypsin is the most commonly used cell dissociation product to obtain primary cells from a tissue sample and to detach cells from a culture vessel.

The use of trypsin for cell dissociation from a tissue sample is mentioned in the particular primary cell culture section (Chapter C). This section describes the protocol for detaching cells from culture vessels.

- Remove and discard the cell culture media
- Wash the culture with 4°C 1 × PBS
- Add trypsin: 2–3 ml for 25 cm² flask; 5 ml for 75 cm² flask; 7–8 ml for 150 cm² flask
- Incubate the flask for 5 mins
- Tap the flask against the palm of your hand to manually detach remaining cells; check under the microscope that all cells are floating
- Neutralise the trypsin by adding double the amount of complete fresh culture media and transfer the solution into centrifuge tubes
- Centrifuge the cells down at 200 g for 10 mins and discard supernatant
- Resuspend the cell pellet, count the cells and replate them at the desired density

Recombinant microbially derived Trypsin (e.g. Gibco™ rProtease) is used in a lesser quantity due to its higher activity. Furthermore, it does not require inactivation or removal, thus eliminating centrifugation and inhibition steps.

Table 3 Suggested wash solutions and trypsin concentrations for dissociating different cell types.

Culture type	Wash solution	Dissociation solution
Strong adherent cells	HBBS or PBS	Trypsin EDTA 0.25%
Epithelial/Endothelial cells	0.5 mM EDTA	Trypsin 0.05%
Multilayer cells	0.05% Trypsin	Trypsin EDTA 0.25%

Note

- Storage: −20°C
- Avoid multiple freeze-thawing, aliquot 5/10 ml samples, or other volumes as required
- Use as little enzyme as possible
- Take special care when opening vials of lyophilized enzymes. Do not inhale the powder!

D.3.4 Other Proteases

A number of other proteases are now commercially available for the purpose of enzymatically dissociating tissue samples. The specificity and application of these are varied and specific protocols for each reader's unique application should be obtained from the literature. Brief descriptions of some of the proteases used by the authors are described here.

Collagenase

Collagenase, as the name suggests, is used to dissociate tissues by breaking down collagen fibers. Since there are several types of collagen based on the composition of the terminal groups at the individual collagen fibrils, different collagenases are also available.

Since collagen is often a major component of the extracellular matrix, collagenase has been the obvious choice used in dissociating tissues for establishing primary cell cultures. This has been shown to be successful with a variety of tissue types. A generic protocol is described here.

- Weigh out the amount of collagenase required to obtain the right concentration (typically 0.1–0.2%, w/v)
- Dissolve collagenase in serum-free DMEM or saline buffer and sterile filter with 0.22 μm filters
- Introduce collagenase solution to minced tissue samples in petri dishes or well plates
- Incubate samples at 37°C with or without shaking, or at 4°C, for the required time period (e.g. dermis digestion can be carried out without shaking at 37°C for 2–4 hrs)

- Filter digested samples and cell suspension mixture through 100 μm nylon sieves to remove large debris
- Centrifuge collected cell suspension at 200 g for 10 mins, and plate accordingly

Note

- Storage: 4°C, lyophilized
- Take special care when opening vials of lyophilized enzymes. Do not inhale the powder!
- Constitute fresh in serum free media or saline buffer
- Sterile-filter before use
- Use as little enzyme as possible

Dispase

Dispase is commonly used for the dissociation of epithelial cells as an entire cell sheet, either from a synthetic culture vessel or from underlying basement membrane. The most common application is the separation of epidermis from the dermis (*C.3.3.1*, pg. 55) and the harvesting of cultured epidermal grafts for clinical applications.

- Weigh out the amount of dispase required to obtain the right concentration
- Dissolve dispase in serum-free DMEM or saline buffer, with agitation, and sterile filter with 0.22 μm filters
- Aspirate media from culture vessel, introduce dispase solution and incubate at 37°C. (The edges of epithelial sheets will start to detach and fold inwards due to contraction, within minutes)

Gently lay a thin layer of gauze or nylon mesh over the detaching cell sheets to allow for easy grafting and to minimise contraction.

Note

- Storage: 4°C, lyophilized
- Take special care when opening vials of lyophilized enzymes. Do not inhale the powder!
- Constitute fresh in serum free media or saline. Dissolves under agitation

- Sterile-filter before use
- Use as little enzyme as possible

Hyaluronidase

Hyaluronidase is a glycanase that degrades hyaluronic acid, a high molecular weight glycosaminoglycan that helps retain water in extracellular matrices. It is used in combination with collagenase in the dissociation of cartilage and liver samples.

Note

- Storage: −20°C, lyophilized
- Take special care when opening vials of lyophilized enzymes. Do not inhale the powder!
- Constitute fresh in serum free media or saline
- Sterile-filter before use
- Use as little enzyme as possible

D.4 Quantitative Cell Proliferation Assays

The measurement of cell proliferation and cell viability has become a key technology to assess the behaviour of cell growth in a culture system. A number of easy and reliable standard assays systems have been developed including: DNA synthesis analysis (PicoGreen® Assay; [³H]–thymidine labelling); cell metabolic assays that are based on the cleavage of tetrazolium salts added to the culture medium (MTS, WST-1); and a non-destructive assay based on a redox indicator (AlamarBlue™).

1. MTS Assay
2. AlamarBlue™ Assay
3. WST-1 Assay
4. [³H]–Thymidine Labelling
5. PicoGreen® Assay

Protect all light-sensitive reagents to prevent photo-degradation

Required Equipment

1. Incubator
2. 96-well Microplate reader with suitable fluorescence/absorbance filters
3. sterile pipette tips
4. 96 well plates
5. PBS, reagent
6. β-counter (Thymidine only)

D.4.1 MTS Cell Proliferation Assay

The assay is based on the cleavage of the MTS tetrazolium compound by cellular enzymes into a coloured formazan product that are soluble in culture medium. The conversion is accomplished by the NADH produced by dehydrogenase in metabolically active cells. The assay is a cell destructive test.

1. Remove media from the culture and rinse gently with PBS
2. Add MTS reagent dissolved in culture media (1:5 ratio) using serum-free media without phenol red

3. Incubate the cells for 3 hrs
4. Vortex the sample with a pipette to ensure mixing
5. Take aliquot and pipette into a 96 well plate
6. Read the absorbance at 490 nm with a reference filter set at 620 nm
7. Use always media with MTS reagent as a control

Note

Storage of MTS reagent: −20°C

D.4.2 AlamarBlue™ Proliferation Assay

The alamarBlue™ assay (Biosource, CA, USA) incorporates a fluorometric/colorimetric growth indicator based on detection of metabolic activity. Specifically, the system incorporates an oxidation-reduction (redox) indicator that both fluoresces and changes colour in response to chemical reduction of growth medium resulting from cell growth.

1. Remove culture medium from wells
2. Rinse with PBS
3. Prepare alamarBlue™ mixture (10% alamarBlue™ stock, 90% colourless culture medium without serum)
4. Add alamarBlue™ to culture wells (100 μl per well for 96-well plates)
5. Incubate for 3 to 5 hours
6. Transfer mixture to 96-well microtiter plates for colorimetric analysis. (100 μl per well)
7. Absorbance is measured at two wavelengths, e.g. 565 nm and 610 nm, and the percentage of reagent reduction is calculated as follows
   a. The absorbance of alamarBlue™ in culture medium is measured at both wavelengths
   b. The absorbance of medium only is also measured at both wavelengths
   c. The absorbance of the medium alone is subtracted from the absorbance of medium plus alamarBlue™ at the higher wavelength. This value is called AO_{HW}

d. The absorbance of the medium alone is subtracted from the absorbance of medium plus alamarBlueTM at the lower wavelength. This value is called AO_{LW}

e. A correction factor R_0 can be calculated from $R_0 = AO_{LW}/AO_{HW}$

f. The percent alamarBlue™ reduced then is expressed as follows:

$$\% \text{ Reduced} = A_{LW} - (A_{HW} \times R_0) \times 100$$

where A_{LW} is the absorbance of sample at the lower wavelength, and A_{HW} is the absorbance of sample at the higher wavelength. A_{LW} and A_{HW} are values after deducting the absorbance of medium alone at the respective wavelengths

8. The second method of monitoring reduction of alamarBlue™ is by measuring the fluorescence at excitation 530–560 nm and emission at 590 nm

While the use of Phenol red has minimal effect on the alamarBlue™ mixture, it may reduce the sensitivity, as well as increase the effect of noise. Use of colourless medium would therefore be prudent. For further details on alamarBlue™ working principles and protocols, please refer to the manufacturer's website at www.biosource.com.

Storage

AlamarBlue™ reagent: 4–8°C

D.4.3 WST–1 Proliferation Assay

WST–1 is designed to be used for the non-radioactive, spectrophotometric quantification of cell growth and viability in proliferation and chemosensitivity assays. It is a colorimetric assay based on the cleavage of the tetrazolium salt WST–1 by mitochondrial dehydrogenases in viable cells. The assay is a cell destructive test.

1. Remove culture medium from wells
2. Rinse with PBS

3. Prepare WST–1 mixture (10% WST–1 stock, 90% phenol red free culture medium with serum)
4. Add WST–1 mixture to culture wells (100 μl per well for 96-well plates)
5. Incubate for 0.5–4 hrs
6. Transfer mixture to 96-well microtiter plates for colorimetric analysis. (100 μl per well)
7. Measure for absorbance at 440 nm against a reference of 600 nm. Shaking to be done at high intensity for 1 min

Storage

WST reagent: −20°C

D.4.4 [³H]–Thymidine Labelling Proliferation Assay

Cells take up [³H]–thymidine during DNA synthesis and repair. Therefore, measuring [³H]–thymidine uptake serves as a semi-quantitative technique for characterising cell proliferation rate. However, [³H]–thymidine labelling data does not reflect the actual number of cells present. An estimated cell number can only be determined by first obtaining a cell number vs. [³H]–thymidine uptake calibration curve for the specific cell type used.

1. Carry out cell culture and maintenance as per normal with [³H]–Thymidine incorporated (3.3 μCi/ml) culture medium, for 24 hrs prior to analysis. (Specific activity range of [³H]–thymidine should be greater than 10 Ci/mmol)
2. Remove culture medium and add 0.1% Triton X–100, in 1 × PBS, to each well/specimen (500 μl for each well in 24-well plate)
3. Leave at room temperature for 15 mins, to permeabilise cell membrane
4. Aliquot 250 μl of permeabilised cell suspension into 4 ml of scintillant and measure radioactivity with a β-scintillation counter

Storage

[³H]–Thymidine reagent: 4–8°C

D.4.5 PicoGreen® DNA Cell Proliferation Assay

PicoGreen® (Molecular Probes Inc., OR, USA) is a reagent for quantifying double-stranded DNA (dsDNA) in solution. The assay is very sensitive and can be used to quantify as little as 25 pg/ml of dsDNA using a standard spectrofluorometer. Fluorescence enhancement of PicoGreen® is measured upon binding dsDNA.

1. Prior to perform the assay, prepare an aqueous working solution of the PicoGreen® by making a 200-fold dilution of the concentrated DMSO stock solution in TE buffer (Tris-HCL/EDTA) and store in a plastic container (e.g. to assay 20 samples in 2 ml volumes, add 100 μl PicoGreen® to 19.9 ml of TE)
2. Prepare reagents for the standard curve: 2 μg/ml stock solution of dsDNA in TE buffer and determine the DNA concentration at an absorbance at 260 nm
3. For sample analysis dilute the experimental DNA in TE to a final volume of 1.0 ml in disposable cuvettes
4. Add 1.0 ml of aqueous working solution of the freshly prepared PicoGreen® reagent to each sample and incubate for 2–5 min at room temperature in the dark
5. Measure the fluorescence using a standard reader (exitation ∼ 480 nm; emission ∼ 520 nm)
6. Subtract fluorescence value of reagent blank from each of the samples and determine DNA concentration from the standard curve

Storage

PicoGreen® reagent: −20°C

D.4.6 Use of Cell Proliferation Assays for High Cell Numbers

It should be noted that when using the above mentioned cell proliferation assays for quantifying high cell numbers, the relationship between cell density and the assay parameter may not be linear. In such cases, intrapolation of the standard curve to determine cell number will not be possible.

As shown in the Fig. D.3, only [^{3}H]–Thymidine Labelling showed good linearity even for cell densities beyond 100,000 per cm^2.

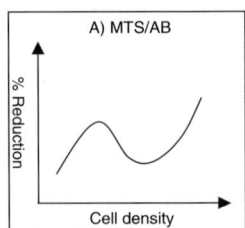 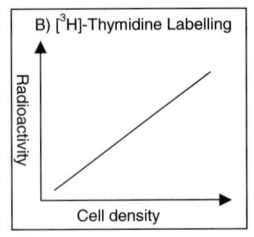 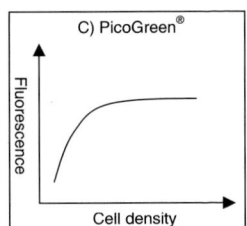

Fig. D.3 Typical trends of using A) MTS/alamarBlue™; B) [³H]-Thymidine Labelling; and C) PicoGreen® for assaying high cell densities in 2D cultures. (Data submitted for publication.)

D.4.7 Colony Forming Unit Fibroblast Assay (CFU–F)

This assay allows the relationship between the number of cells plated and the number of colonies formed to be evaluated. It further allows morphological characterisation of the plated cells and is a useful tool to optimise media or serum conditions for cell cultures.

Unseparated or fractions of bone marrow or peripheral blood cells obtained with the different separation procedures can be assayed. Cells are plated at different densities in T25 tissue culture flasks or into 60 mm petri dishes.

The cells are incubated at 37°C and 5% CO_2 for 14 days. At the end of the culture period, the medium is removed and residual non-adherent cells are removed by washing with phosphate buffered saline

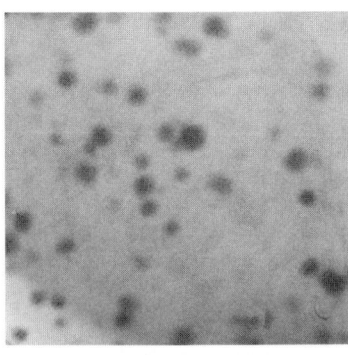 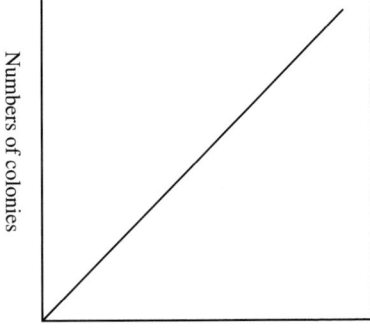

Fig. D.4a Fibroblast-like colony forming units from bone marrow aspirates, stained with Wrights' Giemsa in a T75 flask after 14 days.

Fig. D.4b Corresponding graph showing that generally a linear relationship exists between the number of colonies and the number of cells plated.

pH 7.4 (PBS). The adherent colonies are fixed with concentrated methanol and subsequently stained with Wright's Giemsa (Sigma, St. Louis, USA). A polychromatic stain such as Giemsa stains the nuclei dark blue and the cytoplasm gray-blue. The colonies are scored at ×25 magnification using a phase contrast light microscope.

D.4.8 Growth kinetics of cell populations

Mean generation number *G*

Cells proliferate by cell division. Thus cell populations increase as a power of 2: 1 (2^0), 2 (2^1), 4 (2^2), 8 (2^3), 16 (2^4) ... Therefore we can express the cell proliferation with:

$$N_t = N_0 2^n \Leftrightarrow n = \frac{\log N_t - \log N_0}{0.3}$$

where n is the number of divisions and N_0, N_t are the cell numbers at times 0 and t (hours) respectively. The mean generation time G (hours) is the average time for a cell to divide, thus $G = \dfrac{t}{n}$ and therefore $N_t = N_0 2^{\frac{t}{G}}$

Note

Cell growth characteristics can be classified into three phases: *Lag phase*, *Log phase* and a *Stationary phase*, during which growth usually slows down due to contact inhibition.
The equation only applies when the growth is constant.

Specific growth rate μ

The specific growth rate is based on the natural logarithms for changes in cell concentration or cell numbers at any time of the culture:

$$\mu X = \frac{dX}{dt}$$

or integrated:

$$X_t = X_0 e^{\mu t}$$

where X_t, X_0 are the cell numbers or concentrations at times t (hours) and 0 respectively. μ is the specific growth rate (h^{-1}).

Combination of G and μ

Under the assumption that a doubling in cell number will cause a doubling in concentration, we can write

$$2X_0 = X_0 e^{\mu G}$$
$$\Leftrightarrow 2 = e^{\mu G}$$
$$\Leftrightarrow \ln(2) = \mu G$$
$$\Leftrightarrow 0.693 = \mu G$$
$$\text{or } \mu = \frac{0.693}{G}$$

D.5 Transfection

Transfection of primary human cells refers to the process by which a known DNA molecule is introduced into a recipient cell and is subsequently integrated into the cell's native chromosomal DNA. Transfection is therefore a useful technique to introduce genes into cells when studying gene expression, or simply as a reporter system for the identification or tracking of cells. The common methods of transfection, namely lipofection, electroporation, calcium phosphate precipitation and viral transfection, are described here.

D.5.1 Lipofection

Lipofection can be briefly described as the process of injecting a lipid-DNA complex into cells. Different formulations of lipids are available for this purpose, but they usually consist of mixtures of neutral co-lipids with a cationic lipid to form cationic liposomes, which are then mixed with DNA and introduced to the cells. The cationic liposomes form small (average size 100–400 nm) unilamellar liposomes when formulated in water under optimal conditions. The surface of these liposomes is positively charged and is attracted electrostatically to the phosphate backbone of DNA, as well as to the negatively charged surface of the cell membrane. The negatively charged DNA binds spontaneously to the positively charged liposomes, forming DNA-cationic lipid reagent complexes. It is postulated that entry of the liposome complex into the cell occurs via endocytosis or fusion of the liposome with the plasma membrane. However, it is not clear how the transfected DNA or liposome/DNA complex gains entry into the nucleus.

Efficient transfection using lipofection can only be achieved upon optimising the different parameters involved. A protocol for transfecting primary bone-marrow derived mesenchymal stem cells, used by the authors, is given here as a general guide. Conditions need to be modified and optimised for other cell types.

Transfection with LIPOFECTAMINE™ (Invitrogen Life Technologies)

- Seed 1.5×10^5 cells per well in a 6-well plate, with 2 ml of DMEM with 10% FBS
- Incubate the cells at 37°C in a CO_2 incubator for 24 hrs

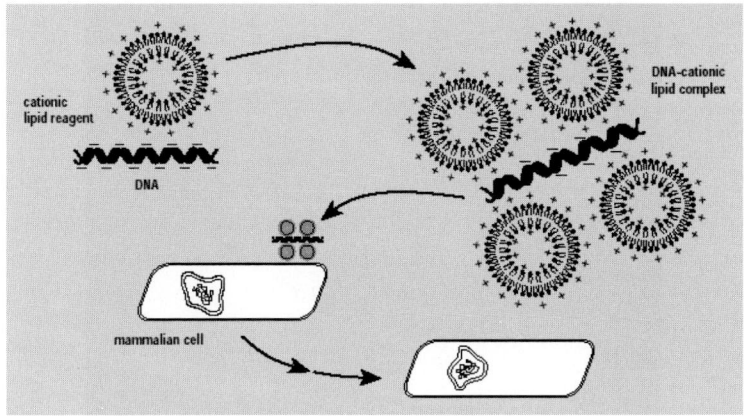

Fig. D.5a Cationic lipid reagent-mediated transfection.

- Prepare the following solutions:
  - * Solution A: For each transfection, dilute 2 μg of DNA into 100 μl OPTI-MEM® I Reduced Serum Medium (Invitrogen Cat. No. 31985–062)
  - * Solution B: For each transfection, dilute 2–16 μl of LIPOFECTAMINE Reagent into 100 μl OPTI-MEM® I Reduced Serum Medium. (Note: Use similar proportions of DNA and LIPOFECTAMINE Reagent in OPTI-MEM® when larger amounts are needed.) Increase the amounts of all reagents in proportion to the surface area when using larger culture vessels
- Combine the two solutions, mix gently and incubate at room temperature for 45 mins. The solution may appear cloudy
- Wash the cells once with 2 ml of serum-free DMEM
- Add 0.8 ml of serum-free DMEM to each tube containing the lipid-DNA complexes
- Do not add antibiotics to media
- Mix gently and overlay the diluted complex solution onto the washed cells
- Incubate the cells for 5 hrs at 37°C in a CO_2 incubator
- Add 1 ml of DMEM with 10% FBS without removing the transfection mixture
- Replace medium at 24 hrs following start of transfection
- Assay cell extracts for gene activity 48 hrs after the start of transfection, depending on cell type and promoter activity

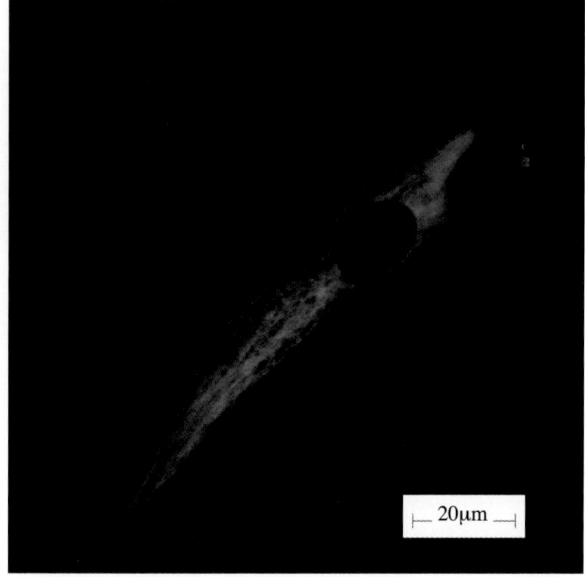

Fig. D.5b **G**reen **F**luorescent **P**rotein (GFP) transfected bone-marrow derived mesenchymal stem cell.

Transfection Optimisation

Careful optimisation of transfection conditions is essential to achieve the highest efficiency possible, at a low cytotoxicity level. The conditions that should be optimised include lipid and DNA concentrations, cell number, and time of exposure of cells to DNA-liposome complexes.

For further protocols and optimisation recommendations using the LIPOFECTAMINE™ reagent, please refer to the supplier's website at http://invitrogen.com.

Requirements

1. DNA plasmid/vector
2. OPTI-MEM® I Reduced Serum Medium
3. LIPOFECTAMINE™ reagent
4. Media: DMEM Supplemented with 10% FBS

D.5.2 Electroporation

Electroporation involves the uptake of DNA into cells, which are subjected to a high-voltage electrical pulse of defined magnitude and length. The mechanism is unclear but it is postulated that transient pores are formed in the cell wall as a result of electroshock, hence enabling macromolecules such as DNA to pass into the cytoplasm. High transfection efficiencies are possible with this method, although it is also associated with high cell death. The maximum voltage and duration of the current pulse are critical in determining the success of transfection.

D.5.3 Calcium Phosphate Precipitation

In this method, insoluble precipitates of calcium phosphate-DNA complexes are formed and introduced to the cells. The precipitates attach to the cells' surfaces and are endocytosed. DNA uptake into the nuclei occurs via an unknown mechanism. The efficiency of transfection with this method is affected by pH, the amount and type of DNA and cell density.

D.5.4 Viral Vectors

Viral vectors are the disabled viruses that have been modified to become replication-incompetent. This technique requires strict precautions to avoid the spread of the virus to the operator and to other members in the laboratory. It should only be used by personnel who have acquired experience in handling viral vectors.

Adenoviral, retroviral, adeno-associated and the haemagglutinating virus of tapain, have all been used for transfection. Adenoviral vectors are very efficient in gene transfer, but the gene is often expressed for a few wells only and is not transmitted to the daughter cells upon cell division.

By contrast, retroviral vector can transfer genes and facilitate its integration to the cell genome so that it is expressed in the daughter cells. However, retroviruses can only transfect dividing cells.

A brief summary of various viral methods used for transfection are outlined in Table 4.

Table 4 Transfection techniques using viral vectors.

Methods	Advantages	Disadvantages
Plasmid DNA	Unlimited transgene size Simple and available technology Episomal location	Low tranfection efficency
Adenovirus	High transfection efficency	Induction of inflammatory reaction
	Limited duration of transgene expression (to prevent over-response)	Non-specific transfection
Retrovirus	Sustained transgene expression	Random nuclear incorporation Applicable only for dividing cells
Adeno-associated virus	Sustained transgene expression	Random nuclear incorporation Complex technology for production
	High transfection efficiency	Limited transgene size

D.6 Reference Guide

D.6.1 Troubleshooting

Table 5 Common problems in cell culture and suggested solutions.

Problems	Possible cause	Suggested solutions
Rapid pH shift in medium	Incorrect carbon dioxide tension	Alter the percentage of CO_2 the incubator based on concentration of sodium bicarbonate medium or switch to CO_2 independent medium
	Overly tight caps on tissue culture flasks	Loosen caps 1/4 turn
	Insufficient bicarbonate buffering	Add HEPES buffer to a final concentration of 10–25 mM
	Bacterial, yeast or fungal contamination	Discard culture and medium or try to decontaminate culture
Precipitate in medium, no change in pH	Residual phosphate left over from detergent washing, which can precipitate powdered media components	Rinse glassware in deionised water several times then autoclave
	Frozen medium	Warm medium to 37°C and swirl to dissolve. If precipitate persists, discard medium
Precipitate in medium, change in pH	Bacterial or fungal contamination	Discard medium and try to decontaminate culture

Table 5 (Cont.)

Problems	Possible cause	Suggested solutions
Primary cell culture is contaminated	Contamination of primary tissue carried over to culture	Wash tissue pieces several times in PBS solution containing higher concentration of antibiotics and antimycotics before culturing
Decrease growth of culture	Change in medium or serum	Compare medium formulations for differences in glucose, amino acids, and other components
		Increase initial cell inoculum
		Adapt cells sequentially to the new medium
	Depletion, absence, or breakdown of essential growth promoting components such as glutamine or growth factors	Remove medium and add fresh medium. Supplement medium with growth promoting components if necessary
	Low level bacterial or fungal contamination	Discard medium and try to decontaminate culture
	Improper storage of reagents	Store sera at −5°C to −20°C and store media at 2°C to 8°C. Store in the dark
	Senescence of finite culture	Discard culture and obtain new cell stock
	Mycoplasma contamination	Segregate culture and test for mycoplasma infection. Clean hood and incubator.

Table 5 (Cont.)

Problems	Possible cause	Suggested solutions
		If culture is contaminated, discard
Death of culture	No CO_2 in incubator	Monitor rate of CO_2 use in incubators and test for leaks. Avoid opening and closing incubator doors unnecessarily
	Temperature fluctuations in the incubator	Monitor temperature of incubator
	Use of fungicide or antibiotics at toxic concentrations	Use less of these reagents
	Cell damage during thawing or cryopreservation	Obtain another aliquot for cells
	Incorrect osmotic pressure in medium	Check osmolarity of complete medium
	Build up of toxic metabolites in medium	Remove and add fresh medium

D.6.2 Cryopreservation and Thawing of Cells

Primary human cells are cryopreserved mainly for storing purposes, as well as to avoid aging and transformation. Before cells are frozen they should be checked for contamination and integrity. Because of the formation of ice crystals, water is removed from the extracellular environment of the cell, causing the intracellular water to migrate outside the cell. Hence the cells suffer from osmotic stress. If too much water remains in the cell, damage due to crystal building in the cells can occur. The extent of the cell damages depends on the cooling rate. Rapid cooling rates effects less osmotic stress but leads to more intracellular ice crystals. Slow cooling rates show the contrary effects of higher osmotic stress and less internal ice. A cooling rate of 1°C per minute and addition of cryoprotectives (DMSO, Glycerol) minimises the damages to the cells.

Notes

- **Storage**: up to 6 month: −80°C; after that transfer to liquid nitrogen, −190°C
- Freeze slowly, thaw fast
- **PROPER TUBE LABELLING**

Freezing

- Detach the cells from the substrate, centrifuge and count them. Assess the viability via trypan blue exclusion
- Resuspend 5×10^6 to 1×10^7 cells/ml in complete media containing 10% DMSO
- Let the suspension equilibrate for at least 15 mins to allow the cryoprotective to penetrate the cells. If the equilibration time exceeds 45 mins, the cryoprotectives may be toxic to the cells
- DMSO is quickly absorbed in the body through the skin and it may transport harmful substances with it. Wear gloves while working with DMSO!
- Label the cryovials properly: preservation date, specimen number, name/identification of cell culturist, cell type, cell number, number of passage, additives and category of danger
- Aliquot into cryogenic storage vials and start the freezing procedure within 5 mins
- Cells are frozen slowly at −1°C/min using a methanol filled cryocontainer, which is placed in a −40°C freezer
- Remove the cells from the cooling unit and place them at an appropriate storage temperature
- Wear insulated gloves, neck shield and a safety mask while working with liquid nitrogen
- Place the storage vials in the allocated boxes in the liquid-nitrogen tank or freezer and make a note in the inventory records cards
- Make sure that the temperature in the freezer or nitrogen tank remains constant

Fig. D.6 Cryocontainer to control temperature drop at −1°C/min.

Thawing

There are two basic methods for thawing up frozen cells: the *direct plating method* and the *centrifugation method*. Both of them are described.

- Find out from the inventory records where your desired samples are located within the liquid-nitrogen tank or freezer to minimise the time for retrieval and risk of warming up the freezer
- Wear insulated gloves, neck shield and a safety mask while working with liquid nitrogen
- Be prepared for leaking or exploding tubes

Direct plating method

- Remove the cryovials from the freezer and quickly thaw them in a 37°C water bath
- Disinfect the external surface of the tubes with 70% ethanol
- Plate the cells directly within complete medium by using 10–20 ml of media per 1 ml of frozen cell suspension
- Change the media after one day

Centrifugation method

- Remove the cryovials from the freezer and quickly thaw them in a 37°C water bath
- Disinfect the external surface of the tubes with 70% ethanol
- Place 1 ml of frozen cell in 20 ml of complete media, gently mix them and centrifuge the cells down at 130 g for 5 mins
- Resuspend the cell pellet in complete media and distribute them into culture flasks

Inventory control

Appropriate record keeping is important to allow rapid and easy retrieval of a specific lot. These records can be maintained on record cards (duplicates!) such as the one shown below.

Name: _____ strain:_____

Growth: Medium _____ Temperature _____ °C Time_____

Special preservation conditions:_____

Lot	Date preserved	Method	Stocks prepared		Purity	Transfer From Original	Ch.	Initials
			Seed	Order				

DMSO: Dimethylsulfoxide   FBS : Fetal bovine serum LN_2 : liquid nitrogen
DW : distilled (sterile) water   HS : Horse serum VLN_2 : Vapor, liquid nitrogen
G : Glycerol   HuS : human serum   Ch : Characterization
S : Sucrose   PF : Protein free
No : no Additive   CD : Chemically defined

D.6.3 Cell Counting with the Haemocytometer

The most direct way of quantifying cell numbers is by counting with a haemocytometer. The haemocytometer is a glass slide with two sets of precision-engineered grid lines on each of two mirror-finished surfaces. The precise manufacturing of the slide ensures that when liquid is introduced between the slide and a cover slip placed over the grid area, a constant amount of liquid (10^{-4} ml) will always be trapped. Counting the number of cells within the grid area will thus yield the cell density of the culture, and subsequently, the total cell number available. Cell counting with a haemocytometer is often done in conjunction with trypan blue exclusion staining. The charged trypan blue molecules do not penetrate into viable cells because of the integrity of the membrane. Non-viable cells, however, allow the free entry of trypan blue, thereby staining the cell interior blue. Therefore, only unstained, viable cells are counted and blue, non-viable cells are excluded. The procedure for cell counting is described below.

- Trypsinize culture in the same way as when subculturing
- After centrifugation, resuspend the cell pellet in 1–2 ml complete media
- Vortex gently to ensure a homogenous single-cell suspension
- Mix 10 µl of the resuspended cell pellet was with 10 µl of 0.4% Trypan Blue in PBS in a micro vial, giving a dilution factor of 2. (Higher dilution factors should be used when higher cell numbers are expected — counting a large number of cells by hand increases the chances of error and thus compromises the accuracy of the count)
- Carefully clean the haemocytometer surface and the cover slip with 70% ethanol and allow it to dry
- Wet sides of the cover slip with reagent grade water and align the cover slip over the haemocytometer
- Dispense the cell-trypan blue mixture onto the haemocytometer, making use of the tapered groove at each end of the mirror surfaces as a guide
- Using a 10 × phase contrast light microscope objective, the number of viable cells within each set of squares (1 mm²) can be counted. (Refer to manufacturer's specifications for grid pattern.)

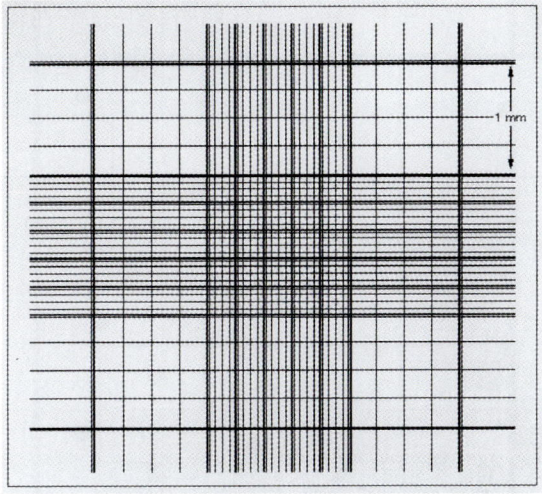

Fig. D.7a An example of grid pattern on a haemocytometer.

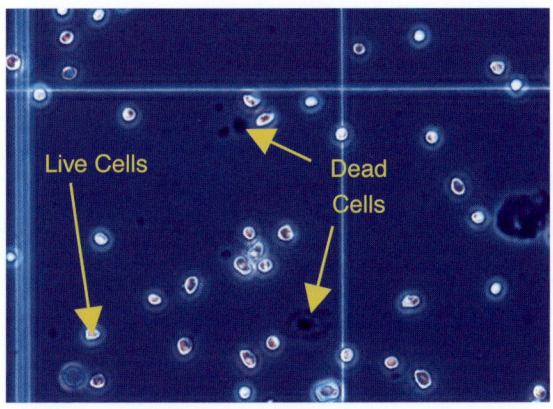

Fig. D.7b Differentiating live and dead cells with trypan blue.

- The number of cells per ml can be counted by:

$$\frac{\text{Cells}}{\text{mL}} = \text{Average of Cells counted} \times \text{dilution factor} \times \text{vol. conversion factor } (10^4)$$

- Multiplying the number of cells per ml by the known volume of cell suspension will give the total number of cells available.

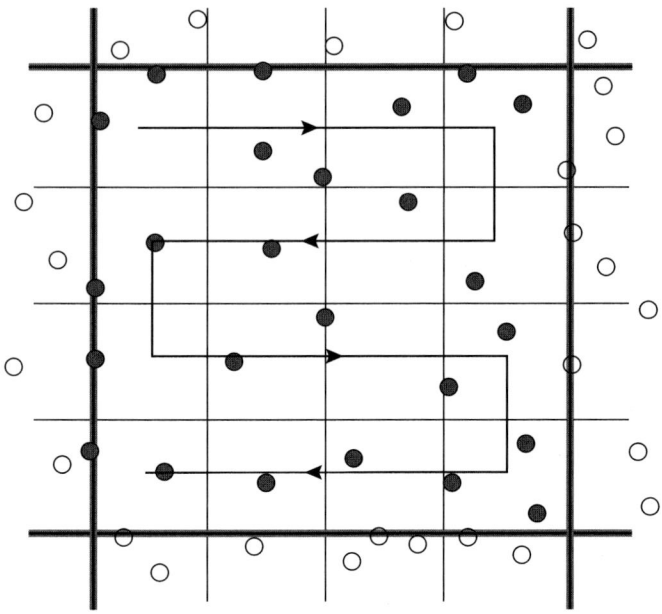

Fig. D.7c Counting cells within the haemocytometer—recommended pattern of counting.

The % viability is the percentage of living cells in the total cell number (stained + unstained cells):

$$\% \text{ Viability} = \frac{\text{Total Cells counted} - \text{stained Cells}}{\text{Total Cells counted}} \times 100$$

Note

- Common errors occur by improper mixing of the cell suspension and/or by allowing the cells to settle down prior to sampling. Avoid the presence of aggregates; this indicates incomplete dissociation.

Notes

Suggested Readings

1. Adams R. L. P., (1990) *Cell Culture for Biochemists: Laboratory Techniques in Biochemistry and Molecular Biology, 2nd* Edition, Elsevier Science Publishers, Amsterdam, The Netherlands

2. Boulton Alan A, Baker Glen B. and Walz Wolfgang, Eds., (1992) *Practical Cell Culture Techniques,* Humana Press Inc, Totowa, NJ

3. Butler M. and Dawson M., Eds., (1992) *Cell Culture: Labfax,* Bios Scientific Publishers, Oxford, England

4. Cohen J. and Wilkin G., Eds., (1996) *Neutral Cell Culture: A Practical Approach,* IRL Press at Oxford University Press, Oxford, England

5. Darling D. C. and Morgan S. J., (1994) *Animal Cells: Culture and Media: Essential Data Series,* John Wiley and Sons Inc, New York, NY

6. Davis J. M., Ed., (1994) *Basic Cell Culture: A Practical Approach,* IRL Press at Oxford University Press, Oxford, England

7. Dixon R. A. and Gonzales R. A., Eds., (1994) *Plant Cell Culture: A Practical Approach, 2nd* Edition, IRL Press, Oxford, England

8. El Oakley R. M., Brand N. J., Burton P. B., McMullen M. C., Adams G. B., Poznansky M. C., Barton P. J. and Yacoub M. H., (1998) Efficiency of a high-titer retroviral vector for gene transfer into skeletal myoblasts. J Thorac Cardiovasc Surg. 115(1):1–8

9. Freshney R. I., Ed., (1992) *Animal Cell Culture: A Practical Approach, 2nd* Edition, IRL Press, Oxford, England

10. Freshney R. I., Ed., (1992) *Culture of Epithelial Cells,* Wiley–Liss, New York, NY

11. Freshney R. I., (2000) *Culture of Animal Cells: A Manual of Basic Techniques (Fourth Edition),* Wiley–Liss Inc, New York, NY

12. Gamborg O. L. and Phillips G. C., (1995) *Plant Cell, Tissue and Organ Culture: Fundamental Methods,* Birkhauser, Boston, MA

13. George E. F., Putock D. J. M. and George H. J., (1987) *Plant Culture Media: Volume 1, Formulations and Uses,* Exegetics, Edington, England

14. George E. F., Putock D. J. M. and George H. J., (1988) *Plant Culture Media: Volume 2, Commentary and Analysis,* Exegetics, Edington, England

15. GIBCO Cell Culture Products, Catalog 2003 Invitrogen Corporation, Carlsbad California, www.invitrogen.com

16. Jakoby W. B. and Pastan I. H., Eds., (1988) *Cell Culture: Methods in Enzymology, Volume 58,* Academic Press, San Diego, CA

17. Jones Gareth E., (1996) *Human Cell Culture Protocols,* Humana Press Inc, Totowa, NJ

18. Klug Christopher A. and Jordan Craig T., Eds., (2002) *Hematopoietic Stem Cell Protocols,* Humana Press, Totowa, New Jersey

19. Koller Manfred R., Palsson Bernhard O. and Masters John R. W., Eds., (2000) *Human Cell Culture: Volume V: Primary Mesenchymal Cells*, Kluwer Academic Publishers, Japan

20. Mah C., Byrne B. J. and Flotte T. R., (2002) Virus-based gene delivery systems. Clin. Pharmacokinet.; 41(12):901–11

21. Toni Lindl, (2002) Zell- und Gewebekultur, Spektrum Akademischer Verlag, Gustav Fischer

22. Martin B. M., (1994) *Tissue Culture Techniques: An Introduction*, Birkhauser, Boston, MA

23. NUNC Cryopreservation Manual, Nalge Nunc International, Naperville, JL, www.nalgenunc.com

24. Pollard J. W. and Walker J. M., Eds., (1997) *Basic Cell Culture Protocols (Second Edition) Volume 75*, Humana Press Inc, Totowa, NJ

25. Potten C., Ed., (1996) *Stem Cells*, Academic Press Inc, San Diego, CA

26. Richmond Jonathan Y., McKinney Robert W., (1999) *Biosafety in Microbiological and Biomedical Laboratories* HHS Publication (CDC) Fourth Edition, NiH

27. Shaw A. J., Ed., (1996) *Epithelial Cell Culture: A Practical Approach,* IRL Press at Oxford University Press, Oxford, England

28. Wise Clare, Ed., (2002) *Epithelial Cell Culture Protocols*, Humana Press, Totowa, New Jersey

Abbreviations

AB/AM	Antibiotics/Antimycotics
BSA	Bovine Serum Albumin
BSS	Balanced Salt Solution
CFU-F	Colony Forming Unit Fibroblasts
DMEM	Dulbecco's Modified Eagle Medium
DMSO	Dimethylsulfoxide
DNA	Deoxyribonucleic acid
ECM	Extracellular Matrix
EDTA	Ethylene Di-Ammine Tetra Acetic Acid
EGF	Epidermal Growth Factor
FACS	Fluorescence Activated Cell Sorting
FBS	Fetal Bovine Serum
HBSS	Hank's Balanced Salt Solution
HEPA	High Efficiency Particulate Air
HEPES	4-(2-Hydroxtethyl)-1-piperazinethansulfonic acid
HUVEC	Human Umbilical Vein Endothelial Cells
LCM	Laser Capture Microdissection
MACS	Magnetic Activated Cell Sorting
MEM	Minimal Essential Medium
MSC	Mesenchymal Stem Cell(s)
MSDS	Material Safety Data Sheet
PBS	Phosphate Buffer Saline
PDL	Periodontal Ligament
RCF	Relative Centrifugal Force

RPM	Revolutions Per Minute
SES	Synthetic Serum Substitutes
STM	Serum-free Medium
SMC	Smooth Muscle Cells
Tris	Tris-(Hydroxymethyl) aminomethane

Milestones in the history of cell culture

1880:   Roux maintained Embryonic chicken in saline solution alive

1900:   Harrison grew cells by hanging drop technique in clotted lymph fluid

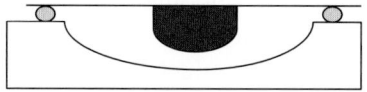

1923:   Carrel introduced aseptic techniques by applying surgical procedures for aseptic handling

1923–31:   Carrel cultured Chicken Embryo Fibroblast during a long term experiment growth-medium: Plasma and tissue homogenate

1940:   First antibiotics became available, decreasing the contamination risk

1940/50:   Polio outbreak, vaccines production with primary monkey kidney cells or human diploid lung fibroblast (first available cell culture product)

1950s: Chemically defined media (Eagle, Earle)
              Advantages: Consistency, sterile, reproducible results

1960:   van Wezel introduced Microcarrier to culture adherent monkey kidney cells in suspension in stirred tank reactors (solved the problem of scaling up)

1965:	Ham described first the use of serum-free medium (SFM)
1970s:	development of DNA-technology made production of recombinant proteins in bacteria, insect and mammal cells possible
1975:	Kohler & Milstein opened the way of unlimited production of monoclonal antibodies by Hybridoma-Technique
1975:	Rheinwald & Green broke ground for therapeutic cell transplantation with primary human keratinocytes to be used in the form of sheet grafts to treat severe burn injuries
1990s:	Autologous and allogenic primary human mesenchymal and hematopoietic cells are transplanted alone or in combination with biomaterials (tissue engineering) to treat a variety of localised and systemic diseases
2000:	Transplantation of genetically modified stem and progenitor cells for tissue regeneration

Current research aims at development of serum-free media, genetically modified cell lines, improvement of culture systems and analysis of the cells and products

Index

Notes

Notes

Notes